Springer Theses

Recognizing Outstanding Ph.D. Research

For further volumes:
http://www.springer.com/series/8790

Aims and Scope

The series "Springer Theses" brings together a selection of the very best Ph.D. theses from around the world and across the physical sciences. Nominated and endorsed by two recognized specialists, each published volume has been selected for its scientific excellence and the high impact of its contents for the pertinent field of research. For greater accessibility to non-specialists, the published versions include an extended introduction, as well as a foreword by the student's supervisor explaining the special relevance of the work for the field. As a whole, the series will provide a valuable resource both for newcomers to the research fields described, and for other scientists seeking detailed background information on special questions. Finally, it provides an accredited documentation of the valuable contributions made by today's younger generation of scientists.

Theses are accepted into the series by invited nomination only and must fulfill all of the following criteria

- They must be written in good English.
- The topic should fall within the confines of Chemistry, Physics, Earth Sciences, Engineering and related interdisciplinary fields such as Materials, Nanoscience, Chemical Engineering, Complex Systems and Biophysics.
- The work reported in the thesis must represent a significant scientific advance.
- If the thesis includes previously published material, permission to reproduce this must be gained from the respective copyright holder.
- They must have been examined and passed during the 12 months prior to nomination.
- Each thesis should include a foreword by the supervisor outlining the significance of its content.
- The theses should have a clearly defined structure including an introduction accessible to scientists not expert in that particular field.

Mika Vesterinen

Z Boson Transverse Momentum Distribution, and ZZ and WZ Production

Measurements Using 7.3–8.6 fb^{-1} of $p\bar{p}$ Collisions at $\sqrt{s} = 1.96$ TeV

Doctoral Thesis accepted by
the University of Manchester, UK

Author
Dr. Mika Vesterinen
University of Manchester
United Kingdom

Supervisor
Prof. Terry Wyatt
University of Manchester
United Kingdom

ISSN 2190-5053
ISBN 978-3-642-30787-4
DOI 10.1007/978-3-642-30788-1
Springer Heidelberg New York Dordrecht London

ISSN 2190-5061 (electronic)
ISBN 978-3-642-30788-1 (eBook)

Library of Congress Control Number: 2012940240

© Springer-Verlag Berlin Heidelberg 2012

This work is subject to copyright. All rights are reserved by the Publisher, whether the whole or part of the material is concerned, specifically the rights of translation, reprinting, reuse of illustrations, recitation, broadcasting, reproduction on microfilms or in any other physical way, and transmission or information storage and retrieval, electronic adaptation, computer software, or by similar or dissimilar methodology now known or hereafter developed. Exempted from this legal reservation are brief excerpts in connection with reviews or scholarly analysis or material supplied specifically for the purpose of being entered and executed on a computer system, for exclusive use by the purchaser of the work. Duplication of this publication or parts thereof is permitted only under the provisions of the Copyright Law of the Publisher's location, in its current version, and permission for use must always be obtained from Springer. Permissions for use may be obtained through RightsLink at the Copyright Clearance Center. Violations are liable to prosecution under the respective Copyright Law.

The use of general descriptive names, registered names, trademarks, service marks, etc. in this publication does not imply, even in the absence of a specific statement, that such names are exempt from the relevant protective laws and regulations and therefore free for general use.

While the advice and information in this book are believed to be true and accurate at the date of publication, neither the authors nor the editors nor the publisher can accept any legal responsibility for any errors or omissions that may be made. The publisher makes no warranty, express or implied, with respect to the material contained herein.

Printed on acid-free paper

Springer is part of Springer Science+Business Media (www.springer.com)

Supervisor's Foreword

This Ph.D. thesis presents three major pieces of new work:

- the development of novel techniques for measuring the transverse momentum of Z bosons;
- the application of these novel techniques to data from the DØ experiment at the Tevatron and a comparison of the measured distributions with state-of-the-art QCD calculations;
- precise measurements of the cross sections for WZ and ZZ vector boson pair production [1].

A novel analysis variable, ϕ_η^*, is proposed [2]. This is sensitive to the same physics as the $p_T^{\ell\ell}$ distribution, but is significantly less susceptible to the effects of detector resolution and efficiency. In particular, since ϕ_η^* depends exclusively on the directions of the two leptons, which are measured with a precision of a milliradian or better, ϕ_η^* is experimentally very well measured compared to any quantities that rely on the momenta of the leptons.

Using 7.3 fb^{-1} of $p\bar{p}$ collisions collected by the DØ detector at the Fermilab Tevatron, the normalized cross section is measured as a function of the variable ϕ_η^* in bins of boson rapidity and in both dielectron and dimuon final states [3]. These measurements are of unprecedented precision and expose deficiencies in the current state-of-the-art QCD predictions for vector boson production at hadron colliders. In addition, these data exclude in hadron–hadron collisions the hypothesis of 'low-x broadening' (This is the idea that the intrinsic transverse momentum of partons inside the proton becomes broader as the momentum fraction x becomes very small).

These new experimental techniques and measurements have stimulated considerable interest in the theoretical and experimental particle physics communities. For example, resummed QCD calculations of the distribution of ϕ_η^* have been made and compared with the above DØ measurements [4]. Measurements of ϕ_η^* are now being extended at the Tevatron to lepton pairs with masses outside the Z resonance region. These techniques have now been picked up by the LHC experiments, where measurements of ϕ_η^* distributions are now being performed.

The measurements of the cross sections for WZ and ZZ vector boson pair production build on earlier measurements by DØ, but use improved analysis techniques and almost the full Tevatron Run II data set. At the time of their publication [1] they represent the world's most precise measurements of the cross sections for WZ and ZZ production. These cross sections are sensitive to anomalous triple-gauge-couplings, and thus probe the electroweak component of the standard model. WZ and ZZ production are two of the smallest cross section processes currently accessible at a hadron collider and represent major sources of background in search channels for Higgs bosons. Understanding these processes is therefore crucial for demonstrating sensitivity to the presence of a standard model Higgs boson at the Tevatron.

Manchester, April 2012   Prof. Terry Wyatt

References

1. DØ Collaboration, V. M. Abazov et al., arXiv:1201.5652v1 [hep-ex] (2012) [submitted to Phys. Rev. D]
2. A. Banfi, S. Redford, M. Vesterinen, P. Waller, T.R. Wyatt, Eur. Phys. J. C **71**, 1600 (2011)
3. DØ Collaboration, V. M. Abazov et al., Phys. Rev. Lett. **106**, 122001 (2011).
4. A. Banfi, M. Dasgupta, S. Marzani, L. Tomlinson, High Energy Phys. **1201**, 044 (2011)

Acknowledgments

First of all I would like to thank my Supervisor, Terry Wyatt, for guiding me through the past 4 years with endless enthusiasm and ingenuity.

Thanks to Fred Loebinger for inspiring me to apply to this program (and also to the MPhys program 4 years earlier), and to Stefan Söldner-Rembold for encouraging me to join the D0 experiment. I am grateful for the tireless IT support (and entertainment) from Sabah Salih while based in Manchester. I would like to thank Mark Owen for helping me to get started with the various D0 frameworks and general coding during my first year. Thanks to all members of the D0 Manchester group and other friends who helped to make my 2 years at Fermilab enjoyable.

Of course I am thankful that my fiancée Joanna Burgess has been so patient and understanding during the last 4 years; particularly the 2 years that I was based at Fermilab. I also thank my parents for always supporting my apparent interest in science from an early age.

Contents

1	Introduction and Theoretical Background	1
	1.1 The Standard Model	1
	1.2 Scattering Amplitudes	3
	1.3 Electroweak Interactions	3
	1.4 Strong Interactions	6
	1.5 The Drell-Yan Process	8
	1.5.1 Factorisation	8
	1.5.2 Higher Order Corrections	9
	1.5.3 Monte Carlo Event Generators	11
	1.5.4 Drell-Yan Transverse Momentum Distribution	13
	1.6 Electroweak Diboson Production	13
	1.6.1 Previous Measurements	14
	References	15
2	Experimental Apparatus	17
	2.1 The Accelerator Chain	17
	2.1.1 The Initial Accelerator Chain	17
	2.1.2 The Main Injector and Antiproton Source	18
	2.1.3 The Tevatron	19
	2.2 The D0 Detector	19
	2.2.1 Inner Tracker	20
	2.2.2 Calorimeter	22
	2.2.3 Muon System	24
	2.2.4 Trigger	25
	References	27
3	Experimental Techniques	29
	3.1 Detector Alignment and Calibration	29
	3.1.1 Calorimeter Calibration	29

	3.2	Particle Reconstruction		30
		3.2.1	Charged Tracks	30
		3.2.2	Electrons and Photons	30
		3.2.3	Muons	32
		3.2.4	Hadronic Taus	33
		3.2.5	IC Electrons	33
		3.2.6	Hadronic Jets	33
		3.2.7	Missing Transverse Energy	34
	3.3	Monte Carlo Simulation	34
	3.4	Unfolding	35
	3.5	Multivariate Classifiers	35
	References	35

4 Electron and Photon Energy Calibration ... 37

	4.1	Introduction	37
	4.2	Dataset	39
	4.3	Correction of the ϕ-Bias	39
	4.4	Energy Calibration	40
		4.4.1	Additional Calibrations	40
		4.4.2	A ϕ_{mod} Dependence Energy Calibration	41
		4.4.3	Shower Shape Dependence	49
		4.4.4	Pseudorapidity Dependence	49
		4.4.5	Fits for the ϕ_{mod} Dependence	49
		4.4.6	Iteration of the Corrections	49
		4.4.7	Resolution Improvements	50
		4.4.8	Energy Dependence	53
	4.5	Monte Carlo Over-Smearing	53
		4.5.1	The Crystal Ball Function	56
		4.5.2	Method to Fit for the Crystal Ball Parameters	59
		4.5.3	Results	62
		4.5.4	Data Versus MC Comparisons	63
	References	63

5 Novel Variables for Studying the Drell-Yan Transverse Momentum ... 65

	5.1	Previous Measurements	65
	5.2	First Idea: the a_T Variable	65
	5.3	Second Idea: Mass Ratios of a_T and $p_T^{\ell\ell}$	67
	5.4	Third Idea: The ϕ_η^* Variable	68
	5.5	Simple Parameterised Detector Simulation	69
	5.6	Scaling Factors	69
	5.7	Experimental Resolution for Dilepton Scattering Angle	70
	5.8	Experimental Resolution for Variables Related to $p_T^{\ell\ell}$	70
	5.9	Acceptance and Efficiency	72

	5.10	Sensitivity to the Physics	74
	5.11	Discussion on the Different Variables.	76
	References	80
6	**Measurement of the Drell-Yan ϕ_η^* Distribution**	81
	6.1	The Observables	81
	6.2	Event Selection	82
		6.2.1 Event Selection Strategy	82
		6.2.2 Data Sample and Skims	84
		6.2.3 Monte Carlo Samples	85
		6.2.4 Common Dielectron and Dimuon Requirements	85
		6.2.5 Dielectron Event Selection	85
		6.2.6 Dimuon Event Selection	86
	6.3	Corrections to the Fully Simulated Monte Carlo Events	89
		6.3.1 Instantaneous Luminosity Profile	91
		6.3.2 Generator Level Physics Re-Weightings	91
		6.3.3 Electron Energy and Muon Momentum Smearing	92
		6.3.4 Track ϕ and η Smearing	92
		6.3.5 Electron Track p_T Smearing	92
		6.3.6 Local Muon p_T Smearing	94
		6.3.7 Trigger and Offline Efficiencies	94
	6.4	Backgrounds	107
	6.5	Comparison of Data with Simulation	110
	6.6	Unfolding	133
		6.6.1 Binning in ϕ_η^*	133
		6.6.2 Bin-by-Bin Unfolding	133
	6.7	Systematic Uncertainties	134
	6.8	Results	135
		6.8.1 Theoretical Predictions	135
		6.8.2 Comparison of Data and ResBos	137
		6.8.3 Fitting for g_2	137
	6.9	Cross Checks	137
		6.9.1 Dielectron Versus Dimuon Comparison	137
		6.9.2 ϕ-Gap Checks	138
		6.9.3 Unfolding Closure Test Using g_2	138
		6.9.4 Data Subset Checks Using g_2	139
	References	139
7	**Measurement of the ZZ and WZ Production Cross Sections**	141
	7.1	Introduction	141
	7.2	Dataset and MC Samples	141
		7.2.1 Dataset	141
		7.2.2 MC Samples	141

	7.3	Dilepton Preselection	142
		7.3.1 Trigger Requirements	143
		7.3.2 Lepton Quality Definitions	143
		7.3.3 $ZZ/\gamma^* \to \nu\bar{\nu}\ell^+\ell^-$ Dilepton Preselection Requirements	143
		7.3.4 $WZ/\gamma^* \to \ell\nu\ell^+\ell^-$ Dilepton Preselection Requirements	145
	7.4	Additional Objects	145
		7.4.1 Jets	145
		7.4.2 Missing Transverse Energy	145
		7.4.3 Additional Leptons	146
	7.5	Corrections to the Simulation	146
	7.6	Comparison of Data and Simulation After Dilepton Selection	149
	7.7	Missing Transverse Momentum Estimators	151
		7.7.1 Construction of the Variables, $\displaystyle{\not{p}'_T\,\not{q}'_T\,\not{q}'_L}$	152
		7.7.2 Calorimeter Recoil	155
		7.7.3 Weighted Combination of $\not{q}'_T$ and $\not{q}'_L$	158
		7.7.4 Comparison of the Discriminating Variables	159
		7.7.5 Performance of the Variables	160
	7.8	Lepton Fake Rate Measurement	164
	7.9	W(+jet) Background Estimation	165
	7.10	$Z/\gamma^* \to \ell^+\ell^-$ Candidate Selection	167
		7.10.1 Signal-Free Control Regions	170
	7.11	$WZ/\gamma^* \to \ell\nu\ell^+\ell^-$ Signal Selection	173
		7.11.1 Selection Cuts	173
		7.11.2 Normalisation of Z/γ^*+jet Backgrounds	179
		7.11.3 Trigger Efficiencies	182
		7.11.4 Signal-Free Control Regions	184
	7.12	$ZZ/\gamma^* \to \nu\bar{\nu}\ell^+\ell^-$ Multivariate Analysis	185
	7.13	Signal Cross Section Measurements	189
	7.14	Systematic Uncertainties	192
	7.15	Results	199
	References		205
8	**Conclusions**		207
Appendix A: Electron and Muon Transfer Functions			211
Appendix B: Systematic Uncertainty Tables by Sub-Channel			215

Chapter 1
Introduction and Theoretical Background

This thesis documents two analyses performed with the large sample of dielectron and dimuon events collected by the D0 detector at the Fermilab Tevatron. The first analysis takes a novel approach to the long studied Drell-Yan transverse momentum distribution, which probes higher order effects in Quantum chromodynamics. The second analysis studies the production of ZZ and WZ in final states with charged leptons and missing transverse momentum.

This chapter introduces the Standard Model of particle physics, and motivates the analyses presented in this thesis. Chapter 2 details the main components of the Tevatron accelerator chain, and the D0 detector. Chapter 3 describes some of the experimental techniques used to reconstruct and identify particles, and to measure their energies/momenta. In Chap. 4, detailed calibrations of electron and photon energies are developed for poorly instrumented regions of the calorimeter. Chapter 5 introduces several novel variables for studying the Drell-Yan transverse momentum distribution. Chapter 6 documents a measurement of the Drell-Yan cross section as a function of one of these variables, ϕ_η^*. Chapter 7 documents a measurement of the ZZ and WZ production cross sections. Finally, Chap. 8 presents the conclusions drawn from the work in this thesis.

1.1 The Standard Model

The Standard Model (SM) of particle physics has been extremely successful in describing the interactions of high energy particles, up to the energies that have been probed by experiments so far—roughly one TeV. This theory was largely based on the work of Glashow [1], Weinberg [2] and Salam [3], and earned them the 1979 Nobel prize.

Matter is understood in terms of twelve elementary spin-$\frac{1}{2}$ fermions (the quarks and leptons), as listed in Table 1.1, with each having an anti-matter equivalent. Interactions between these fermions are mediated by the spin-1 vector bosons; the photon (γ) of the electromagnetic force, the $W^\pm$ and Z bosons of the weak force, and the

Table 1.1 The elementary particles in the SM, all of which have been observed experimentally except for the Higgs boson

Fermions				Bosons
Quarks:	$\begin{pmatrix} u \\ d \end{pmatrix}$	$\begin{pmatrix} c \\ s \end{pmatrix}$	$\begin{pmatrix} t \\ b \end{pmatrix}$	$\gamma, W^{\pm}, Z, g$
Leptons:	$\begin{pmatrix} e \\ \nu_e \end{pmatrix}$	$\begin{pmatrix} \mu \\ \nu_\mu \end{pmatrix}$	$\begin{pmatrix} \tau \\ \nu_\tau \end{pmatrix}$	H

gluon (g) of the strong force. Certain interactions between the bosons themselves are also predicted. An additional spin-0 scalar Higgs field (H) provides a mechanism to give masses to the weak bosons, whilst allowing the photon to remain massless. The Higgs field also provides masses for the fermions. Apart from the Higgs, all of these particles have been observed experimentally. A quantum theory that includes gravity remains a theoretical dream for now.

Much of this structure results from requiring that the SM Lagrangian is invariant under *local gauge transformations*, $\psi \to \psi e^{i\phi(x)}$, where *local* specifies that the phase ϕ depends on the space-time coordinates x. This condition requires the existence of the gauge bosons and their interactions with the fermions. The interactions of the SM particles are described by the set of gauge groups

$$SU(3)_C \otimes SU(2)_L \otimes U(1)_Y,$$

where $SU(3)_C$, $SU(2)_L$ and $U(1)_Y$ are the gauge groups of colour, weak isospin, and weak hypercharge respectively. Each of these gauge groups will be described in more detail in the sections to follow.

Despite the remarkable success of the SM (we shall see results of some of the most stringent experimental tests in Sect. 1.3), it is still assumed to be a low energy effective theory. Firstly, the SM does not contain a natural candidate for the dark matter required by observations in astrophysics. Secondly, the SM is unable to account for the matter anti-matter asymmetry of the universe. Thirdly, the Higgs boson mass would acquire large loop corrections from whatever physics lies between the Electroweak scale and the Planck scale.[1] A popular extension to the SM that attempts to solve the problems mentioned above is Supersymmetry, where each SM particle has a superpartner with a different spin. Fermions have scalar (sfermion) superparters and bosons have fermion (gaugino) superpartners. Reference [4] provides an excellent overview of Supersymmetry. An alternative is to introduce extra dimensions through which gravity is allowed to propagate, but confining the SM particles to a 4D "brane" [5].

---

[1] At the Planck scale quantum effects on gravity become significant. The Planck scale is defined as $\Lambda_{\text{Plank}} = (8\pi G_N)^{-1/2} \approx 10^{19}$ GeV, where G_N is the gravitational constant.

1.2 Scattering Amplitudes

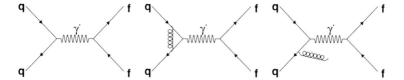

Fig. 1.1 *Left* tree level Feynman diagrams for quark anti-quark annihilation to a pair of fermions through a virtual photon. *Centre* similar diagram with a virtual gluon exchange between the incoming quark and anti-quark. *Right* similar diagram with a real gluon emission off one of the quarks

1.2 Scattering Amplitudes

In the limit that the relevant coupling constants are small (the perturbative regime), cross sections for particle interactions can be calculated using the method of Feynman diagrams. Figure 1.1 (left subfigure) shows a Feynman diagram for the annihilation of a quark and antiquark into a pair of fermions via the exchange of a virtual photon. A quantum mechanical amplitude for such a diagram is calculated by assigning a multiplicative factor for each line and vertex. Figure 1.1 also shows two (of the many) additional diagrams which appear at higher orders in Quantum chromodynamics, which will be described in Sect. 1.4. The central diagram includes a *virtual* or loop correction, and the right hand diagram includes a *real* or radiative correction. The accuracy of cross section calculations depends on the order of diagrams considered. Typically, the coupling constant associated with higher order corrections is small enough that the problem can be treated *perturbatively*, i.e., the series of successive orders converges.

1.3 Electroweak Interactions

Electroweak interactions are described by the $SU(2)_L$ gauge group of weak isospin and the $U(1)_Y$ gauge group of weak hypercharge. The subscript L indicates a distinction between left- and right-handed fermions.[2] Left(right) handed fermions act as doublets(singlets) under the $SU(2)_L$ gauge group. The third component of the weak isospin, T_3, is a conserved quantum number in SM interactions. Thus from now on, "weak isospin" refers to T_3. Weak hypercharge, Y_W, is defined as $Y_W = 2(Q - T_3)$, where Q is the electric charge. Table 1.2 lists the values of T_3 for the left- and right-handed fermions (and anti-fermions). The gauge bosons associated with the $SU(2)_L$ and $U(1)_Y$ groups are the $W^0/W^\pm$ and the B respectively. The B and W^0 are rotated to give the physical (mass eigenstates) γ and Z bosons:

$$\begin{pmatrix} Z \\ \gamma \end{pmatrix} = \begin{pmatrix} \cos\theta_W & \sin\theta_W \\ -\sin\theta_W & \cos\theta_W \end{pmatrix} \begin{pmatrix} W^0 \\ B \end{pmatrix},$$

---

[2] A right-(left-) handed particle has its spin and momentum vector pointing in the same (opposite) direction.

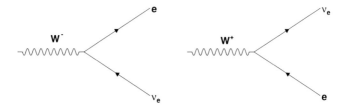

Fig. 1.2 Feynman diagrams for the decays, $W^+ \to e^+\nu_e$ and $W^- \to e^-\bar{\nu}_e$

where θ_W is the weak mixing angle that relates the coupling constants associated with the $SU(2)_L$ (g) and $U(1)_Y$ (g') gauge groups:

$$\tan\theta_W = \frac{g'}{g}.$$

This angle also relates the masses of the $W^\pm$ and the Z. At tree level, the SM predicts the following relation

$$\cos\theta_W = \frac{m_W}{m_Z}.$$

The masses of the W and Z boson are generated through the Higgs mechanism [6, 7]. An additional complex scalar, Higgs, field has a symmetry that is spontaneously broken at the electroweak scale. This produces additional degrees of freedom that generate gauge invariant mass terms for the W and Z whilst allowing the photon to remain massless. An additional scalar particle, the Higgs boson is also generated. The Higgs boson is the only fundamental particle in the SM that has not been observed directly in experiment, and its discovery is currently one of the main goals in the field.

The allowed charged current interactions are: the conversion of a neutrino to a charged lepton of the *same generation* or vice versa, and the conversion of an up-type quark to *any* (kinematically allowed) down-type quark or vice versa. Figure 1.2 shows Feynman diagrams for the decays, $W^+ \to e^+\nu_e$ and $W^- \to e^-\bar{\nu}_e$. The mixing between different quark flavours is governed by a unitary 3×3 matrix—the so-called Cabibbo-Kobayashi-Maskawa matrix [8, 9].

Neutral current interactions are mediated by the photon and the Z boson. The coupling constant associated with a photon-fermion vertex is eQ_f, where e is the electromagnetic coupling constant, and Q_f is the charge of the fermion. The Z boson couples to left- and right-handed fermions as

$$g^f_{(L,R)} = T^f_{3(L,R)} - Q_f \sin^2\theta_W.$$

Since the left- and right-handed fermions have different T_3 (see Table 1.2), they acquire different couplings to the Z. Traditionally, the Z-fermion coupling has been

1.3 Electroweak Interactions

Table 1.2 The weak isospin (T_3^f) and electric charge (Q_f) for the different fermion (f) and antifermion ($\bar{f}$) types. The three up- and down-type quarks are represented by q_u and q_d respectively. The three charged and neutral leptons are represented by q_u and q_d respectively. The subscripts L and R denote left- and right-handed fermions

Fermion	$T_{3,L}^f (T_{3,L}^{\bar{f}})$	$T_{3,R}^f (T_{3,R}^{\bar{f}})$	$Q_f (Q_{\bar{f}})$
$\begin{pmatrix} q_u \\ q_d \end{pmatrix}$	$+\frac{1}{2}(0)$ $-\frac{1}{2}(0)$	$0(-\frac{1}{2})$ $0(+\frac{1}{2})$	$+\frac{2}{3}(-\frac{2}{3})$ $-\frac{1}{3}(+\frac{1}{3})$
$\begin{pmatrix} l \\ \nu_l \end{pmatrix}$	$-\frac{1}{2}(0)$ $+\frac{1}{2}(0)$	$0(+\frac{1}{2})$ $0(-\frac{1}{2})$	$+1(-1)$ $0(0)$

written in a vector minus axial-vector (V-A) form, with a vector coupling constant $c_V = g_L + g_R$ and an axial coupling constant $c_A = g_L - g_R$.

The structure of the electroweak theory has been verified to high precision at the LEP and SLC e^+e^- colliders. The Tevatron has also played an important role in discovering the top quark, and in determining the masses of the top quark and W boson. Since the SM relates the different EW parameters, it is able to predict the value of many parameters and observables, given a limited number of inputs. Figure 1.3 compares the experimental measurements of various EW observables with the SM predictions. The agreement is remarkable and consolidates the SM.

An obvious missing piece is the, as of yet, unobserved Higgs boson. The SM is able to predict the mass of the Higgs, m_H, since it introduces significant loop corrections to the masses of top-quark and the W boson as illustrated in Fig. 1.4. Figure 1.5 shows the current bounds imposed on m_H by the measurements of m_t and m_W. Currently the bounds on m_H are more limited by the precision of m_W than the precision of m_t. Thus an improved measurement of m_W is an important goal of the Tevatron and LHC experiments. Figure 1.6 compares the experimental measurements of m_W performed so far. Combination of precision electroweak data (excluding direct Higgs searches) excludes[3] $m_H > 161$ GeV [10]. The regions $m_H < 114$ GeV and $156 < m_H < 177$ GeV have been excluded in direct searches at LEP [11], and the Tevatron [12]. A recent combination of direct searches by the ATLAS and CMS Collaborations using up to 2.3 fb^{-1} of pp collisions has excluded the region $141 < m_H < 476$ GeV [13]. If the SM Higgs exists in the theoretically preferred low mass region, the LHC experiments are likely to see evidence within the next year or so.

An important feature of the electroweak sector is the *non-abelian* structure of the $SU(2)_L \otimes U(1)_Y$ gauge group. This implies certain interactions between the bosons themselves. These triple-gauge-couplings play an important role in boson pair production.

---

[3] All exclusions that are quoted in this section are at 95 % C.L.

Fig. 1.3 Loop corrections to the W boson propagator from (*left*) the top quark and (*right*) the Higgs boson

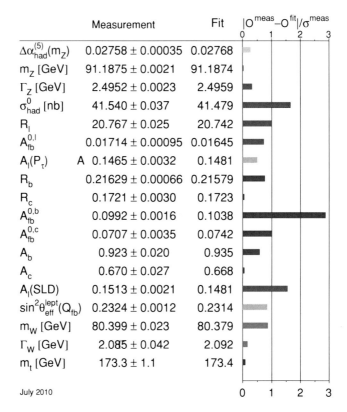

Fig. 1.4 Comparison of EW measurements from experiments at LEP, SLC, and the Tevatron, with a global fit (from Ref. [14]). The *horizontal bars* indicate the number of standard deviations by which the direct measurement differs from the global fit

1.4 Strong Interactions

The strong force is described by the $SU(3)_C$ gauge group of Quantum chromodynamics (QCD), where rotations in 3-dimensional colour space are mediated by eight unitary 3×3 matrices—the Gell-Mann matrices.

1.4 Strong Interactions

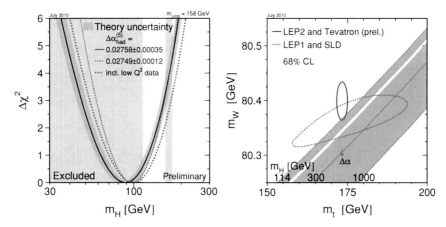

Fig. 1.5 The *left hand figure* shows the SM preferred region for m_H, where the *yellow shaded* regions are directly excluded by experimental searches. It should be noted that the direct searches from LHC have recently excluded the region $m_H > 141\,\text{GeV}$ [13]. The *right hand figure* shows the experimental measurement contours for m_W and m_t. The *green bands* show the SM predictions for different values of m_H. Both figures are from Ref. [14]

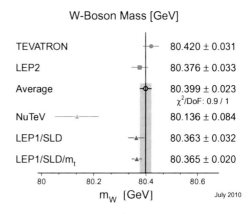

Fig. 1.6 Comparison of direct and indirect experimental measurements of m_W (from Ref. [14])

The coupling constant associated with colour exchange vertices is usually denoted g_s, though it is often more convenient to work in terms of $\alpha_s = g_s^2/4\pi$. An important feature of QCD is the running of α_s with energy scale. For n_f colours, and one-loop precision, the evolution of α_s with energy scale, Q, is given by

$$\frac{d\alpha_s}{d\log Q} = -\left(11 - \frac{2n_f}{3}\right)\frac{\alpha_s^2}{2\pi}. \quad (1.1)$$

Figure 1.7 shows various measurements of α_s at different values of energy scale. The yellow band is a prediction from QCD, calculated at four-loop precision, having

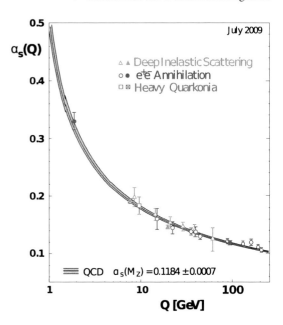

Fig. 1.7 Various measurements of α_s over a range of energy scales (from Ref. [15]). The *yellow band* is the prediction from QCD

constrained to the average of the measurements at $Q = m_Z$. It can be seen that the predicted evolution of α_s with scale agrees well with experimental observation.

At small energies (or correspondingly large distances) α_s is large, leading to the confinement of quarks to colourless hadrons; either baryons containing three quarks of different colour, or mesons containing a quark anti-quark pair of the same colour (anti-colour for the anti-quark). The confinement scale of QCD, Λ_{QCD}, is naturally around the mass of the lightest hadron (the pion) which is approximately 100 MeV.

High energy quarks and gluons (typically referred to as "partons") produced in particle scatterings readily radiate additional partons. This "showering" stops once the parton energies reach Λ_{QCD}, at which point they become confined to colourless hadrons. High energy partons therefore end up as so called "jets" of collimated hadrons.

1.5 The Drell-Yan Process

The Drell-Yan process is production of a lepton pair in hadron–hadron collisions via a virtual photon (γ^*), Z, or W [16].

1.5.1 Factorisation

At high enough energies such that α_s becomes sufficiently small (roughly 10^{-1} at electroweak scales), quark-antiquark scattering cross sections can be calculated

1.5 The Drell-Yan Process

perturbatively. However quarks are confined to colourless hadrons which are inherently non-perturbative objects.

Luckily, calculations of hadron–hadron scattering cross sections can be *factorised* into the following general form:

$$\sigma = \sum \int_{x_a=0}^{x_a=1} \int_{x_b=0}^{x_b=1} dx_a dx_b f_a(x_a, Q^2) f_b(x_b, Q^2) \sigma_{\text{hard}}(Q^2),$$

where the sum runs over parton species, and x_a and x_b are the fractions of the parent hadron momenta carried by the interacting partons. The functions f_a and f_b are the so called parton distribution functions (PDFs), which parameterise the probability to extract a parton with a certain fraction, x, of the hadron momentum. The PDFs depend on the parton species (gluon, u-quark, d-quark etc.) and the scale of the process, Q^2. Calculation of the hard scattering cross section, σ_{hard}, can be done perturbatively. The Drell-Yan processes is special—thus far it is the only hadron–hadron scattering process for which factorisation has been demonstrated mathematically [17].

The PDFs can be constrained by various hadron–hadron and hadron–lepton scattering data over a range of x and Q^2 values. References [18] and [19] provide details on some of the most recent fits by the CTEQ and MSTW groups respectively. Extrapolation to different Q^2 values is governed by the DGLAP [20–22] equations of QCD.

1.5.2 Higher Order Corrections

Neglecting (for now at least) any intrinsic transverse motion of the partons within the hadrons, the Drell-Yan process should, at lowest order, produce a dilepton system with zero momentum transverse to the beam direction, p_T. However, at next-to-leading order in the strong coupling, an initial state quark can radiate a gluon (see right hand of Fig. 1.1), thus generating a non zero p_T. This (real correction) diagram is actually "infra-red" divergent when the radiated gluon is soft (i.e. low momentum) and/or collinear with the parent parton. The loop (virtual correction) diagram (middle of Fig. 1.1), is both "infrared" and "ultraviolet" divergent since one can integrate up to infinite loop momenta. All is not lost, since the divergences of the real and virtual diagrams actually cancel, at least when calculating the inclusive cross section [23–25].

Prediction of the p_T distribution of the dilepton system poses a problem, since, at a given order in α_s, the real and virtual diagrams populate different phase space. This results in incomplete cancellation of the divergences. leading-order (LO) in the p_T distribution only includes the left hand diagram in Fig. 1.1. Next-to-LO (NLO) includes the right hand diagram corresponding to one power of α_s, and the centre diagram corresponding to α_s^2. Although the loop diagram contains two powers of

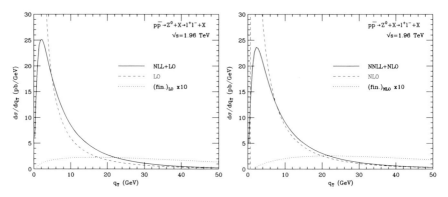

Fig. 1.8 Predictions of the Z/γ^* p_T (here given the symbol q_T) distribution from Ref. [26], up to (*left*) LO and LO+NLL, and (*right*) NLO and NLO+NNLL

α_s, it is included at NLO through interference with the L0 diagram. Figure 1.8 shows calculations from Ref. [26] of the Z/γ^* p_T distribution using different levels of accuracy. The NLO prediction is clearly divergent at low p_T. At any finite order in α_s, the coefficient of the α_s^n term contains $1/p_T^2$ multiplied by a series of logs, $\ln^m(Q/p_T)$, $m = 0, 1, .., 2n - 1$. These logs become large at low p_T and spoil the convergence of the series.

A complete prescription for the Drell-Yan process at low p_T was introduced by Collins et al. [17], based on the ingredients provided by Refs. [27–31]. The expansion in powers of α_s is reorganised in terms of the logarithms, which can then be recognised as the Taylor expansion of an exponential. The large logarithms are thus exponentiated or "resummed" to all orders in α_s. This gives a finite result which can be matched[4] to fixed order calculations at larger p_T. The accuracy of the resummation depends on the order of the logs considered. Leading-log (LL) includes the (leading) $\ln^{2n-1}$ logs. Higher logarithmic accuracies include the sub-leading logs. Figure 1.8 shows that the resummation, either at next-to-LL (NLL) or next-to-NLL (NNLL), yields a finite cross section over the entire p_T range. The resummation actually needs to be performed in impact parameter ($b \sim 1/p_T$) space, such that momentum conservation in multiple parton emission can be factorised.

For $p_T < \Lambda_{\rm QCD}$, we encounter a further problem as the largeness of α_s renders QCD non-perturbative. Instead, non-perturbative (NP) functions must be determined from fits to experimental data. Various forms for the NP functions have been suggested in the literature, for example that of Brock, Landry, Nadolsky and Yuan (BLNY) [32]:

$$W(b, Q, Q_0, x_1, x_2) = \exp\left(\left[-g_1 - g_2 \ln(\frac{Q}{2Q_0}) + g_1 g_3 \ln(100 x_1 x_2)\right] b^2\right).$$

---

[4] The matching procedure must ensure that perturbative terms are not double counted by the fixed order and resummed calculations.

1.5 The Drell-Yan Process

They performed a global fit to Run I Tevatron Z/γ^* data, and lower Q^2 Drell-Yan data from various fixed target experiments. Fixing $Q_0 = 1.6\,\text{GeV}$, their fit found $g_1 = 0.21^{+0.01}_{-0.01}\,\text{GeV}^2$, $g_2 = 0.68^{+0.01}_{-0.02}\,\text{GeV}^2$, and $g_3 = 0.60^{+0.05}_{-0.04}$.

The Collins-Soper-Sterman (CSS) formalism has been implemented in the MC program RESBOS[33] with the above form factor. Figure 1.9 shows the RESBOS prediction of the p_T distribution in three bins of the dilepton rapidity, y, defined as

$$y = \frac{1}{2}\ln\left(\frac{E+p_z}{E-p_z}\right),$$

where E is the dilepton energy, and p_z is the dilepton momentum along the beam direction. For Z/γ^* production at the Tevatron, $Q \sim 90\,\text{GeV}$, and $x_1, x_2 \sim 10^{-3} - 10^{-1}$. In this case, the p_T distribution is most sensitive to the g_2 parameter, as can be seen in Fig. 1.9—a larger value of g_2 corresponds to a broader p_T distribution.

A particularly interesting part of the NP form factor is the x dependence. Semi inclusive deep inelastic scattering data from HERA [34, 35] indicates a broadening of the form factor at low values of x [36]. For Z/γ^* production at the Tevatron, the boson rapidity is related to the x of the two partons:

$$x_{1,2} = \frac{Q}{\sqrt{s}}e^{\pm y},$$

where $\sqrt{s}$ is the hadron–hadron centre of mass energy of the collider. Large values of boson rapidity correspond to one parton with small x and one parton with large x. The "small-x broadening" [36] would widen the predicted p_T distribution at large values of $|y|$, as shown in Fig. 1.9. The effect becomes significant for $|y| > 2$, corresponding to one parton with $x < 10^{-2}$. At the LHC, *inclusive* production of W, Z, and Higgs bosons involves partons with similarly small values of x. The small-x broadening would therefore have a dramatic effect on the p_T spectra in these processes at the LHC [37]. Early measurements from the ATLAS [38] and CMS [39] are in reasonable agreement with predictions that do not include such effects.

1.5.3 Monte Carlo Event Generators

Much of the hadron collider physics analysis program relies on Monte Carlo event generators to predict kinematic distributions and rates for signal and background processes. So far, the state-of-the-art resummation programs (e.g. RESBOS) have only been able to predict distributions for the final state leptons, rather than the full particle content.

The programs PYTHIA [40] and HERWIG++ [41] match leading-order (LO) matrix elements to leading-log (LL) parton showers, which evolve high energy partons down to some predefined cut-off scale by radiating additional partons. Models for

Fig. 1.9 Predictions of the p_T distribution in $p\bar{p}$ collisions at $\sqrt{s} = 1.96$ TeV from RESBOS for three different dilepton rapidity ranges. The different curves correspond to (*black*) default $g_2 = 0.68\,\text{GeV}^2$, (*dashed red*) $g_2 = 0.46$ GeV^2, and (*dashed blue*) $g_2 = 0.68\,\text{GeV}^2$ but including the small-x effect [36]

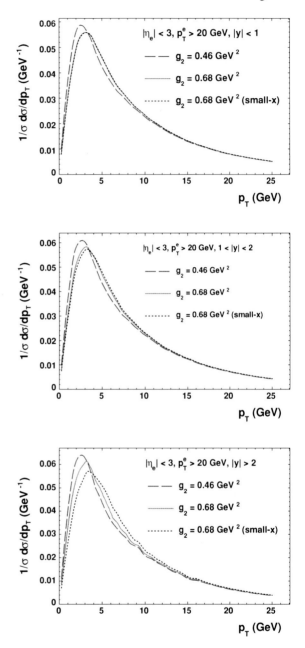

hadronisation and the underlying event[5] are also implemented. These models require

---

[5] The underlying event is activity that is not associated with the hard parton-parton scattering, e.g., the hadron remnants or multiple parton scattering.

tuning to describe data over a wide range of processes. In addition, significant physics that is missing at LO/LL must be absorbed by model parameters. More recently, methods to match next-to-LO matrix elements to parton showers have been introduced as implemented in the POWHEG [42, 43] and MC@NLO [44] programs. An alternative is to match higher multiplicity LO matrix elements to parton showers, as implemented in the ALPGEN [45] and SHERPA [46] programs.

1.5.4 Drell-Yan Transverse Momentum Distribution

Similar physics applies to the p_T distribution of any hadron collider physics process. Uncertainties in predicting this production mechanism degrade the sensitivity of searches for the Higgs boson, and beyond SM physics. Also, the important m_W measurement described earlier relies on an accurate model of the kinematics in W boson production. Current m_W measurements at the Tevatron are based on the RESBOS program [33] using the BLNY NP form factor, and are sensitive to the value of g_2. Precise measurements of the Drell-Yan p_T distribution can be used to verify the accuracy of the state-of-the-art calculations, constrain any non-perturbative effects, and tune simpler MC event generator models. Chapter 5 will introduce alternative observables (a_T and ϕ_η^*) that are sensitive to the Drell-Yan p_T, but are more optimal from an experimental point of view. State-of-the-art QCD predictions for the a_T distribution have been calculated by Banfi, Dasgupta and Duran Delgado [47]. This calculation was recently extended by Banfi, Dasgupta, and Marzani to the ϕ_η^* variable [48].

1.6 Electroweak Diboson Production

Figures 1.10 and 1.11 show the leading-order Feynman diagrams for the production of ZZ/γ^* and WZ/γ^* respectively, with decays into the specific channels studied in this thesis. In the case of WZ/γ^* production, the left hand diagram includes the WWZ/γ^* vertex, which results from the non-abelian nature of the electroweak theory.

The diboson production cross sections are some of the lowest that can be studied in existing Tevatron and LHC datasets. Figure 1.12 compares the measured and predicted cross sections for various SM Electroweak processes, including diboson production. Measured cross sections that differ from the SM predictions would be an indication of new physics. Currently, the measurement uncertainties for the ZZ/γ^* and WZ/γ^* cross sections are significantly larger than the uncertainties on the predictions.

Diboson processes are a major background in searches for both new physics and the SM Higgs boson. For example, $ZZ/\gamma^* \to \nu\bar{\nu}b\bar{b}$ is a background in the search for $ZH \to \nu\bar{\nu}b\bar{b}$. It is therefore important to verify the accuracy of the production models used in these searches.

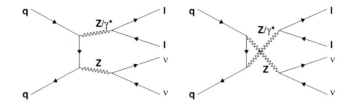

Fig. 1.10 Leading-order Feynman diagrams for the process, $p\bar{p} \rightarrow ZZ \rightarrow l^+l^-\nu\nu$, within the SM

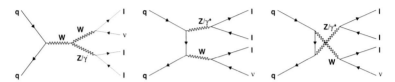

Fig. 1.11 Leading-order Feynman diagrams for the process, $p\bar{p} \rightarrow WZ \rightarrow l^+l^-l\nu$, within the SM

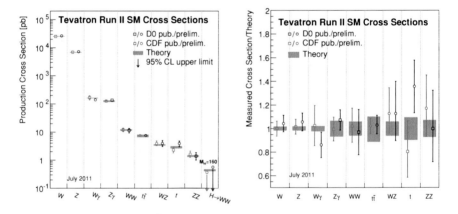

Fig. 1.12 Comparison of measured and predicted cross sections for various SM Electroweak processes

1.6.1 Previous Measurements

A previous D0 analysis of the $ZZ/\gamma^* \rightarrow \nu\bar{\nu}\ell^+\ell^-$ process with $2.7\,\text{fb}^{-1}$[49], measured a production cross section of $\sigma(p\bar{p} \rightarrow ZZ) = 2.01 \pm 0.93(\text{stat}) \pm 0.29(\text{syst})$ pb.[6] The CDF Collaboration has released a preliminary study of this channel with $5.6\,\text{fb}^{-1}$[50] with a measured cross section of $\sigma(p\bar{p} \rightarrow ZZ) = 1.45^{+0.92}_{-0.87}$ pb. D0 recently published an analysis in the $\ell^+\ell^-\ell^+\ell^-$ decay channel with $6.4\,\text{fb}^{-1}$[51],

---

[6] This measurement was translated from the observable ZZ/γ^* cross section to a "pure" ZZ cross section.

yielding a measurement of $\sigma(p\bar{p} \to ZZ) = 1.40^{+0.43}_{-0.37}(\text{stat}) \pm 0.14(\text{syst})$ pb, when combined with the earlier analysis in the $\ell^+\ell^-\nu\bar{\nu}$ channel.

A study of the $WZ/\gamma^* \to \ell\nu\ell^+\ell^-$ process was performed by CDF using 1.1 fb^{-1} of data [52]. This analysis measured a cross section of $\sigma(p\bar{p} \to WZ) = 5.0^{+1.8}_{-1.6}$ pb. A D0 analysis with 4.1 fb^{-1} in the same channel measured a cross section of $3.90^{+1.06}_{-0.90}$ pb [53]. CDF has recently released a preliminary result in this channel with 7.1 fb^{-1} [54], yielding a measured cross section of 3.9 ± 0.8 pb.

References

1. S.L. Glashow, Partial-symmetries of weak interactions. Nucl. Phys. **22**, 579–588 (1961)
2. S. Weinberg, A model of leptons. Phys. Rev. Lett. **19**, 1264–1266 (1967)
3. A. Salam, *Weak and electromagnetic interactions*, ed. by N. Svartholm. Proceedings of the 8th Nobel Symposium on Elementary Particle Theory, Relativistic Groups and Analyticity, Stockholm, Sweden, 1968, pp. 367–377, 1969
4. S.P. Martin, arXiv:hep-ph/9709356v5 (2008)
5. N. Arkani-Hamed, S. Dimopoulos, G. Dvali, The hierarchy problem and new dimensions at a millimeter. Phys. Lett. B **429**(3–4), 263–272 (1998)
6. P. Higgs, Broken symmetries and the masses of gauge bosons. Phys. Rev. Lett. **13**, 508–509 (1964)
7. F. Englert, R. Brout, Broken symmetry and the mass of gauge vector mesons. Phys. Rev. Lett. **13**, 321–322 (1964)
8. N. Cabibbo, Unitary symmetry and leptonic decays. Phys. Rev. Lett. **10**, 531–532 (1963)
9. M. Kobayashi, T. Maskawa, CP violation in the renormalizable theory of weak interaction. Prog. Theor. Phys. **49**, 652–657 (1973)
10. ALEPH Collaboration, CDF Collaboration, D0 Collaboration, DELPHI Collaboration, L3 Collaboration, OPAL Collaboration, SLD Collaboration, LEP Electroweak Working Group, Tevatron Electroweak Working Group, SLD electroweak heavy flavour groups, arXiv:1012.2367[hep-ex] (2011) (unpublished)
11. R. Barate et al., Search for the standard model higgs boson at lep. Phys. Lett. B **565**, 61–75 (2003)
12. TEVNPH (Tevatron New Phenomena and Higgs Working Group), CDF and D0 Collaboration, arXiv:1107.5518[hep-ex] (2011) (unpublished)
13. ATLAS and CMS Collaborations, ATLAS-CONF-2011-157, CMS PAS HIG-11-023 (2011) (unpublished)
14. J. Alcaraz et al., Latest plots from the LEP Electroweak Working Group, July, 2011
15. S. Bethke, The 2009 world average of α_s. Eur. Phys. J. C **64**, 689–703 (2009)
16. Sidney D. Drell, Tung-Mow Yan, Massive lepton-pair production in hadron–hadron collisions at high energies. Phys. Rev. Lett. **25**, 316–320 (1970)
17. J. Collins, D. Soper, G. Sterman, Transverse momentum distribution in Drell-Yan pair and W and Z boson production. Nucl. Phys. B **250**, 199–224 (1985)
18. H. Lai et al., New parton distributions for collider physics. Phys. Rev. D **82**, 074024 (2010)
19. A. Martin, W. Stirling, R. Thorne, G. Watt, Parton distributions for the lhc. Eur. Phys. J. C **63**, 189–285 (2009)
20. V.N. Gribov, L.N. Lipatov, Sov. J. Nucl. Phys. **15**, 438 (1972)
21. G. Altarelli, G. Parisi, Nucl. Phys. B **126**, 298 (1977)
22. Y.L. Dokshitzer, Sov. Phys. JETP **46**, 641 (1977)
23. T. Kinoshita, J. Math. Phys. **3**, 650 (1962)
24. T.D. Lee, M. Nauenberg, Phys. Rev. **133**, 1549 (1964)

25. N. Nakanishi, Progress Theoret. Phys. **19**, 159 (1958)
26. G. Bozzi, S. Catani, G. Ferrera, D. de Florian, M. Grazzini. Production of Drell-Yan lepton pairs in hadron collisions: Transverse-momentum resummation at next-to-next-to-leading logarithmic accuracy. Phys. Lett. B **696**, 207 (2011)
27. Y.L. Dokshitzer, D. Diakonov, S.I. Troyan, Phys. Lett. B **79**, 269–272 (1978)
28. G. Altarelli, G. Parisi, R. Petronzio, Phys. Lett. B **76**, 356 (1978)
29. G. Parisi, R. Petronzio, Nucl. Phys. B **154**, 427 (1979)
30. C.T.H. Davies, W.J. Stirling, Nucl. Phys. B **244**, 337 (1984)
31. C.T.H. Davies, B.R. Webber, W.J. Stirling, Nucl. Phys. B **256**, 413 (1985)
32. F. Landry, R. Brock, P.N. Nadolsky, C.-P. Yuan, Fermilab tevatron run-1 z boson data and the collins-soper-sterman resummation formalism. Phys. Rev. D **67**, 073016 (2003)
33. C. Balazs, C.-P. Yuan, We use the CP version of the code and grid files. Phys. Rev. D **56**, 5558–5583 (1997)
34. C. Adloff, Eur. Phys. J. C **12**, 595 (2000)
35. J. Breitweg et al., Phys. Lett. B **481**, 199 (2000)
36. P. Nadolsky, D.R. Stump, C.-P. Yuan, Phys. Rev. D **64**, 114011 (2001)
37. S. Berge, P. Nadolsky, F. Olness, C.-P. Yuan, Phys. Rev. D **72**, 033015 (2005)
38. G. Aad et al., Measurement of the transverse momentum distribution of bosons in protonproton collisions at with the atlas detector. Phys. Lett. B **705**(5), 415–434 (2011)
39. S. Chatrchyan et al., CMS Collaboration, arXiv:1110.4973[hep-ex] (2011)
40. T. Sjostrand, Comp. Phys. Commun. **135**, 238 (2001)
41. M. Bahr, Eur. Phys. J. C **58**(4), 639–707 (2008)
42. P. Nason, J. High Energy Phys. **0411**, 40 (2004)
43. S. Frixione, P. Nason, C. Oleari, J. High Energy Phys. **0711**, 70 (2007)
44. S. Frixione, B.R. Webber, J. High Energy Phys. **0206**, 29 (2002)
45. M.L. Mangano, J. High Energy Phys. **07**, 001 (2003)
46. T. Gleisberg et al., Event generation with SHERPA 1.1. J. High Energy Phys. **02**, 007 (2009)
47. A. Banfi, M. Dasgupta, R.M. Duran Delgado, The a_t distribution of the z boson at hadron colliders. J. High Energy Phys. **0912**, 022 (2009)
48. Andrea Banfi, Mrinal Dasgupta, Simone Marzani, QCD predictions for new variables to study dilepton transverse momenta at hadron colliders. Phys. Lett. B **701**, 75–81 (2011)
49. V.M. Abazov et al., Phys. Rev. D **78**, 072002 (2008)
50. T. Aaltonen et al., CDF note 10358, CDF (2010)
51. V.M. Abazov et al., Phys. Rev. D **84**, 011103 (2011)
52. T. Aaltonen et al., Phys. Rev. Lett. **98**, 161801 (2007)
53. V.M. Abazov et al., Phys. Lett. B **695**, 67 (2011)
54. T. Aaltonen et al., CDF note 10176, CDF (2011)

Chapter 2
Experimental Apparatus

2.1 The Accelerator Chain

Figure 2.1 shows a schematic of the Tevatron and other components of the accelerator chain. Reference [1] provides an excellent introduction to the Tevatron and injector accelerator chain.

The Tevatron is a superconducting synchrotron, proton anti-proton collider, with a circumference of roughly 6 km. Up to 1995 (Run I), the Tevatron collided at a centre of mass energy of $\sqrt{s} = 1.8$ TeV, with a design luminosity, $\mathcal{L}^{\text{inst}}$, of 10^{30} cm^{-2} s^{-1}. For a bunch crossing frequency of f, luminosity is defined as

$$\mathcal{L}^{\text{inst}} = \frac{f N_p N_{\bar{p}}}{4\pi \sigma_p \sigma_{\bar{p}}},$$

where $N_p(N_{\bar{p}})$ and σ_p ($\sigma_{\bar{p}}$) are the number of particles per bunch and the transverse width of the proton (antiproton) beam respectively. Significant upgrades were made to the Tevatron and to the experiments before the start Run II (in 2001). The centre of mass energy was increased to $\sqrt{s} = 1.96$ TeV and, thanks to the addition of the main injector and recycler, the design luminosity was increased to 200×10^{30} cm^{-2}s^{-1}. In fact, luminosities have now exceeded 400×10^{30} cm^{-2}s^{-1}. Details of the Run II accelerator upgrade can be found in Ref. [2].

2.1.1 The Initial Accelerator Chain

The accelerator chain starts with a humble bottle of hydrogen gas. A 750 kV Cockroft-Walton generator ionises the H$_2$ molecules into H$^-$ ions. These ions are then accelerated through a 150 m linac up to an energy of 400 MeV. At this point, a thin carbon foil strips the H$^-$ ions of their electrons, and the resulting protons proceed into the

M. Vesterinen, *Z Boson Transverse Momentum Distribution, and ZZ and WZ Production*, Springer Theses, DOI: 10.1007/978-3-642-30788-1_2,
© Springer-Verlag Berlin Heidelberg 2012

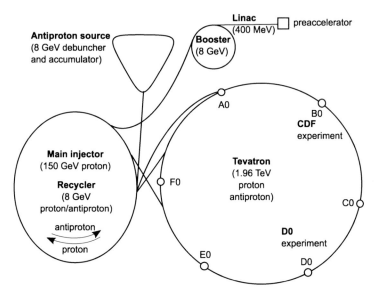

Fig. 2.1 Schematic diagram of the Tevatron and accelerator supply chain

booster ring. The booster is a circular synchrotron which accelerates the protons up to 8 GeV.

2.1.2 The Main Injector and Antiproton Source

Protons are extracted at 8 GeV from the booster into the main injector, which accelerates them to 150 GeV, ready for injection into the Tevatron. Antiprotons, $\bar{p}$s, are produced by bombarding a Ni target with 120 GeV protons from the main injector, every 1.5 s. The angular spread of the produced particles is reduced through a lithium lens. A pulsed magnet mass-charge spectrometer selects the small fraction ($\sim 10^{-5}$) of desired $\bar{p}$s from the large number of particles produced in the p-Ni collisions.

At this point, the $\bar{p}$s have a very large spread in energy, but a relatively small spread in time. The debuncher is a triangular accelerator, which effectively has an energy dependent phase. Higher energy $\bar{p}$s travel further around the debuncher, and thus see a different phase of the accelerating RF cavity, compared to lower energy $\bar{p}$s. Thus, the $\bar{p}$s are gradually "debunched", meaning that the temporal spread is increased whilst the energy spread is decreased. After ~ 1.5 s, the pulse of $\bar{p}$s is moved to the accumulator, making space for the next incoming pulse. The accumulator "stacks" $\bar{p}$s from the debuncher over a period of a few hours. During this time, the energy spread and transverse size of the beam are reduced using the method of stochastic cooling [3]. The $\bar{p}$s are periodically transferred from the accumulator to the recycler, which sits on top of the main injector, and also uses the stochastic cooling method.

2.1 The Accelerator Chain

The $\bar{p}$s are injected into the main injector which ramps their energy up to 150 GeV ready for injection into the Tevatron.

2.1.3 The Tevatron

Protons and antiprotons are transferred from the main injector to the Tevatron at 150 GeV. The Tevatron is a superconducting synchrotron accelerator which increases the beam energies to 980 GeV. Particles at the edges of the beam are removed by collimators. All magnets within the Tevatron lattice are cooled by liquid helium. Dipole magnets operate at 4.2 T, and maintain the circular orbit. There are 36 proton and 36 antiproton bunches, which cross every 396 ns. The D0 detector was originally designed for the much larger 3.6 μs bunch spacing of Run I. Whilst allowing a significant increase in instantaneous luminosity, the shorter Run II bunch spacing required significant upgrades to the trigger and readout system, and still causes challenges in calorimetry as we shall see in Chap. 4. Close to the CDF and D0 detectors, superconducting "low beta" quadrapole magnets focus the beams towards a narrow collision region at the detector centres. Periods of beam collisions are referred to as "stores", typically lasting around 10 h. The transition time in between stores is roughly 1–2 h.

2.2 The D0 Detector

Figure 2.2 shows the layout of the D0 detector. A detailed description of the detector can be found in Ref. [4]. D0 uses a right hand cylindrical coordinate system, where the z axis is along the proton direction, the y axis is upward, and the x axis points to the centre of the accelerator. The azimuthal and polar angles are denoted as ϕ and θ respectively. A vector pointing towards the centre of the accelerator defines $\phi = 0$, and ϕ increases anti-clockwise, such that $\phi = \pi/4$ points upwards. The pseudorapidity, η, is defined as $\eta = -\ln[\tan(\theta/2)]$. The region of small $|\eta|$ ($|\eta| < 1$, say) is typically referred to as "central", and the region of larger $|\eta|$ as "forward". In many situations it is more appropriate to use the detector pseudorapidity, η_{det}, which is defined by a line connecting the passage of a particle in a particular detector subsystem and the centre of the detector.

The following sections describe in more detail the specific detector subsystems which are used extensively in this work. Particular mention is made to regions in ϕ which are poorly instrumented: the central calorimeter module boundaries, and the muon system octant boundaries. The reduced lepton identification efficiencies in these regions requires careful treatment in the measurement of the Drell-Yan ϕ_η^* distribution described in Chap. 6.

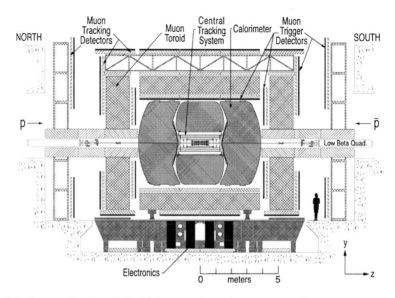

Fig. 2.2 Cross section through the D0 detector, viewed from the centre of the collider ($r - z$ plane). The major sub detector systems are labelled (from Ref. [4])

2.2.1 Inner Tracker

Figure 2.3 shows the layout of the inner tracker region of the detector. The two main components are the silicon microstrip tracker (SMT) and the central fibre tracker (CFT). The beam pipe is made from beryllium, with an outer diameter of 38.1 mm, length of 2.37 m, and thickness of 0.508 mm.

Figure 2.4 shows the layout of the SMT. A detailed description of the SMT can be found in Ref. [5]. The central region in z is covered by 6 barrels, interspersed with 6 so-called F-disks. The forward and backward regions each have a set of three F-disks and additional so-called H-disks. This (disk/barrel) design ensures that tracks generally meet a perpendicular surface, given a luminous region with a roughly a 25 cm wide Gaussian profile along the z axis.

It was decided to upgrade the SMT after Run IIa of the Tevatron ($\int \mathcal{L}dt \approx 1$ fb^{-1}). The closest layer to the beam pipe (layer-1) was projected to suffer serious performance degradation due to radiation damage. A more "radiation hard" layer, the so-called layer-0, was inserted very close to the beam line, adding an additional tracking point, and maintaining vertexing capability whilst the layer-1 performance inevitably degrades. Details of the layer-0 upgrade can be found in Ref. [6].

Figure 2.5 shows a view of the CFT in the $r - \phi$ plane. The CFT consists of 8 cylinders, each of which contains one doublet layer of fibres in the beam direction and a second doublet layer at a stereo angle in ϕ of $+3$ or $-3°$. The outermost layer of the CFT covers the region up to $|\eta| \approx 1.7$. Wavelength shifting optical fibres transport light to visible light photon counters (VLPCs) located in cryostats

2.2 The D0 Detector

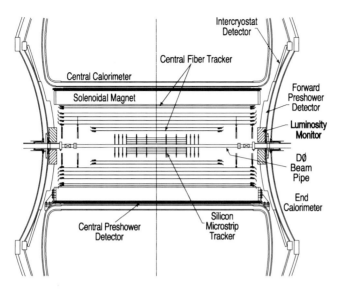

Fig. 2.3 View of the inner tracker region of the detector in the $r - z$ plane (from Ref. [4])

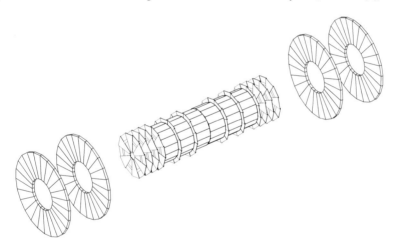

Fig. 2.4 Layout of the SMT (from Ref. [5]). The barrels are coloured black, F-disks in blue, and H-disks in red (Color online)

underneath the detector. The region $\phi \approx \pi/4$ requires longer fibres to reach the VLPC cryostats. We shall later see that this introduces a significant ϕ dependence of the track reconstruction efficiency.

A solenoidal magnet provides a field strength of roughly 2 T, and is designed to have a uniform field throughout its bore. In order to minimise the effect of detector charge asymmetries (e.g., due to mis-alignments) on physics analyses, the magnet polarity is reversed at regular intervals.

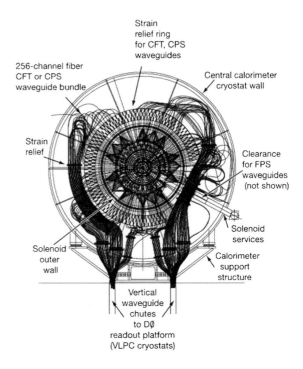

Fig. 2.5 End view of the central calorimeter and CFT waveguides that transport the scintillator light down to the VLPC cryostats underneath the detector (from Ref. [4])

Additional preshower detectors, the central preshower detector (CPS) and forward preshower detector (FPS) cover the regions $|\eta| < 1.3$ and $1.5 < |\eta| < 2.5$ respectively. The CPS sits between the solenoid and the central calorimeter. The FPS is mounted on the outside of the end cap calorimeter cryostat. Both preshower detectors share the same VLPC and readout technology as the CFT.

2.2.2 Calorimeter

Figure 2.6 shows an isometric view of the calorimeter, which is contained in three separate cryostats. The central calorimeter (CC) covers the region up to $|\eta| \approx 1.1$, and the two end caps (ECs) extend the coverage to $|\eta| \approx 4$. There are three layers: electromagnetic (EM), fine hadronic (FH) and coarse hadronic (CH). The region $1.1 < |\eta| < 1.5$ has little or no EM coverage, and limited hadronic coverage. This region is additionally instrumented with scintillator tiles, called the inter-cryostat detector (ICD).

Figure 2.7 shows the layout of a single calorimeter cell. Each cell alternates layers of uranium absorber (iron in the CH layers) and copper resistive plates with a high voltage (roughly 2 kV) in between. The gaps between plates are filled with liquid argon. An electron takes on average 400 ns to drift across the gap [7]. The average

2.2 The D0 Detector

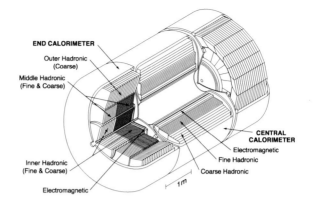

Fig. 2.6 Cut away view of the calorimeter (from Ref. [4]). The three cryostats, and the different layers (*electromagnetic, fine and coarse hadronic*) are labelled

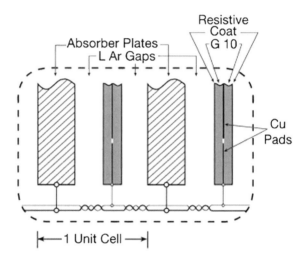

Fig. 2.7 Layout of a single calorimeter cell (from Ref. [4]). The absorber plates are grounded and the resistive plates are connected to a positive high voltage (roughly 2 kV)

electron drift time across the gap is comparable to the 396 ns bunch spacing in Run II of the Tevatron. Charge integration is therefore a significant problem.

The calorimeter cells are arranged such that "towers" of constant η can be drawn projecting from the centre of the detector through the centre of the cells. This "pseudo-projective" geometry is shown for one quarter of the calorimeter in Fig. 2.8. Cells have an angular size of $\Delta\eta = \Delta\phi = 0.1$, apart from the extreme forward region, where the cells are larger in $\Delta\eta$. In addition, cells in the third EM layer, corresponding to the shower maximum for electrons and photons, have a smaller size of $\Delta\eta = \Delta\phi = 0.05$ in order to improve the spatial resolution of electromagnetic cluster centroids.

There are four EM layers in both the CC and EC, with increasing absorber thickness (1.4, 2.0, 6.0, 9.8 X_0 in the CC and 1.6, 2.9 7.9, 9.3 X_0 in the EC, where X_0 is one radiation length). In the EM layers, the absorber is almost 100 % pure depleted uranium. The FH layers use roughly 6 mm thick uranium-niobium alloy, and the CH layers use 46.5 mm thick copper (CC) or brass (EC) plates.

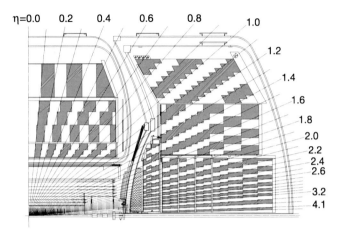

Fig. 2.8 Cross section of one quarter of the calorimeter, viewed from the side (from Ref. [4]). Notice the pseudo-projective geometry with constant cell size in $\Delta\eta$ (apart from the extreme forward region of the EC)

Cells are arranged in modules, one of which is shown in Fig. 2.9. In the CC, there are 32 EM modules, and 16 FH and CH modules, as shown in Fig. 2.10. The EM energy response is degraded near to the poorly instrumented CC module boundaries (ϕ-gaps). This is more problematic with the shorter Run II bunch spacing since the electric field is non-uniform near to the ϕ-gaps, leading to significant losses in charge integration. In Chap. 4, we determine corrections for these energy losses and improve the accuracy with which they are modelled in the MC simulation.

2.2.3 Muon System

Figure 2.11 shows an exploded view of the muon system, which is instrumented with wire drift tubes and plastic scintillator detectors. The wire chambers of the central ($|\eta| < 1$) muon system are called proportional drift tubes, and those of the forward ($1 < |\eta| < 2$) muon system are called mini drift tubes, due to their relative compact size. A 1.8 T toroidal magnet allows a momentum measurement that is independent of the central tracking system, and is used in the level-1 trigger (more later). The toroid polarity is regularly reversed such that roughly equal size datasets are recorded for the four combinations of solenoid and toroid polarities. The muon system is divided into three layers: the A layer sits inside the toroid, and the B and C layers sit outside the toroid. The central region underneath the detector is relatively poorly instrumented with muon detectors (particularly the A layer and C layer scintillators) to make space for the detector support structure and much of the readout hardware, including the VLPC cryostats described in Sect. 2.2.1. Both the central and forward muon systems

2.2 The D0 Detector

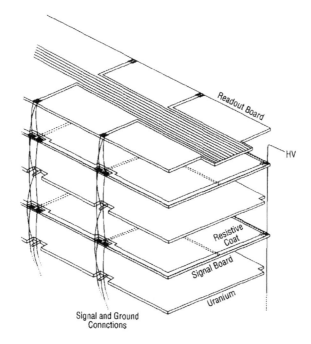

Fig. 2.9 View of a CC module, with the absorber, signal and readout boards labelled (from Ref. [4])

are divided into eight so-called "octants" in ϕ, with limited instrumentation in the boundaries separating the octants.

2.2.4 Trigger

A three-level trigger system is required to reduce the event rate from the roughly 1.7 MHz bunch crossing frequency to around 200 Hz at which events can be recorded. The two analyses presented in this thesis mostly rely on triggers that require one or two high transverse momentum, p_T, electrons or muons.

Level-1 Trigger

The level-1 trigger is based on fast hardware decisions and reduces the rate to around 2 kHz.

- **Central track trigger (CTT)** The CTT looks for predefined patterns in the CFT, CPS, and FPS, consistent with the passage of a charged particle. A limited number of equations (96) for different p_T thresholds can be programmed into the readout.

Fig. 2.10 End view of the CC, showing the module structure (from Ref. [4])

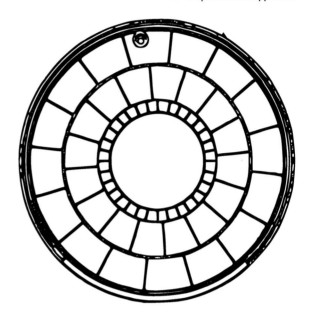

- **Calorimeter trigger (L1Cal)** The L1Cal looks for individual towers above predefined E_T thresholds, for efficient triggering on high p_T electrons, photons and jets.
- **Muon trigger (L1Muon)** The L1Muon reconstructs track stubs with both scintillator and wire hits, or connects wire hits with seed central tracks from the CTT.

Level-2 Trigger

The level-2 trigger refines the energy and momentum measurements from level-1, and combines information from different detector sub-systems to make global decisions. Seed tracks from the L1CTT are fed to the L2 silicon track trigger (STT), which improves the momentum resolution of the CTT tracks. The STT is also able to reconstruct the impact parameter of tracks, which allows tagging of b-flavoured hadrons. A global L2 trigger combines information from all detector subsystems, to calculate object correlations, e.g., invariant masses of particle pairs. The event rate is reduced from 2 kHz to around 1 kHz.

Level-3 Trigger

The level-3 trigger fully reconstructs events using simplified versions of the offline software, and reduces the rate to around 200 Hz, at which events are recorded.

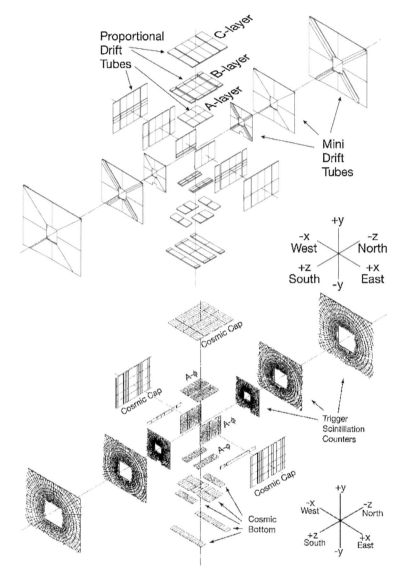

Fig. 2.11 Layout of the wire (*upper*) and scintillator (*lower*) muon detectors (from Ref. [4])

References

1. J. Thompson, Introduction to Colliding Beams at Fermilab. Fermilab Technical Memo 1909, Fermilab, (1994)
2. G. Jackson, Fermilab Recycler Ring Technical Design Report. Fermilab Technical Memo 1991, Fermilab, (1996)
3. J. Marriner, Stochastic cooling overview. Nucl. Instr. Meth. Phys. Res. A **532**(1–2), 11–18 (2004)

4. V.M. Abazov et al., The upgraded D0 detector. Nucl. Instr. Meth. Phys. Res. A **565**(2), 463–537 (2006)
5. S.N. Ahmed, Nucl. Instr. Meth. Phys. Res. A **634**, 8–46 (2011)
6. R. Angstadt et al., Nucl. Instrum. Meth. Phys. Res. A **622**, 298–310 (2010)
7. S. Abachi et al., Nucl. Instr. Meth. A **338** 185 (1994)

Chapter 3
Experimental Techniques

3.1 Detector Alignment and Calibration

The detector must undergo regular calibrations and alignments. The inner tracker is aligned using collision data, and cosmic ray muon events that are recorded whilst the solenoid is switched off.

3.1.1 Calorimeter Calibration

Calibration of the calorimeter can be broadly divided into two parts: readout electronics and cell gains. The readout electronics are calibrated by injecting known amounts of charge. This so called "pulser" based calibration takes place in between Tevatron stores. The gain calibration takes place less frequently (once a year or so—and typically after Tevatron shutdown periods), and is performed in two steps:

- ϕ-**intercalibration** Since the Tevatron beams are unpolarised, the average particle energies and rates should be uniform in ϕ. Event samples are taken with special triggers that apply very loose E_T requirements. For a given ring in η the calibration constants for each cell in ϕ are varied until the energies and rates are uniform in ϕ. This is done separately for the EM and hadronic layers. The ϕ-intercalibration is documented in Refs. [1] (hadronic) and [2] (EM).
- η-**intercalibration** A final step uses dijet events for the hadronic calorimeter, and $Z/\gamma^* \to e^+e^-$ events for the EM calorimeter, to equalise the response as a function of η. For the EM calorimeter, the absolute energy scale is effectively calibrated using the value of m_Z measured at LEP [3]. Prior to calibrating the gain, corrections need to be applied for energy losses in the material upstream of the calorimeter. These are determined from Monte Carlo simulations. An energy scale correction that corrects hadronic jets back to particle level jets is determined separately and is described later in this chapter.

3.2 Particle Reconstruction

3.2.1 Charged Tracks

Charged particle tracks are reconstructed from hits in the SMT and CFT. They are defined by their curvature (q/p_T), angle with respect to a plane that is perpendicular to the bending field (tan λ), and coordinates at the distance of closest approach to the beam pipe. The curvature resolution is approximated by

$$\delta(q/p_T) = A + B/p_T,$$

where the first term relates to the tracker length, magnetic field strength, and the spatial resolution of individual hits. The second term represents the effect of multiple Coulomb scattering.

3.2.2 Electrons and Photons

Electromagnetic clusters (EM clusters) are reconstructed from isolated deposits of energy in the EM layers of the calorimeter. A clustering algorithm firstly looks for individual cells above a certain threshold. Such seed cells are combined with neighbouring cells within a cone of $\Delta \mathcal{R} < 0.2$, where $(\Delta \mathcal{R})^2 = (\Delta \eta)^2 + (\Delta \phi)^2$. EM-clusters can also be matched to charged particle tracks, in which case they are considered as electron candidates. Otherwise they are considered as photon candidates.[1] The energy resolution of a sampling calorimeter can be written in the following general form [4]:

$$\frac{\Delta E}{E} = C \oplus S E^{-\frac{1}{2}} \oplus N E^{-1}$$

where $\oplus$ represents addition in quadrature, and the three terms are understood as follows:

- The constant term ($=C$) is due to spatial and temporal non-uniformities in the calorimeter response.
- The sampling term ($\propto E^{-\frac{1}{2}}$) is due to the statistical nature of a sampling calorimeter. The energy measurement is essentially a statistical inference of a true energy based on an observed number of ionisation electrons.

---

[1] Actually, many physics analyses consider electrons without track matches in order to gain acceptance at the expense of increased instrumental backgrounds. The analyses in this thesis only consider electrons that are matched to central tracks. Non track matched electrons are however used in the measurement of tracking efficiencies.

3.2 Particle Reconstruction

- The final term ($\propto E^{-1}$) is due to noise from the readout electronics and the low level radioactivity in the uranium absorber.

Hadronic jets have a very low probability to be mis-identified as isolated electron signatures. However, given the huge rate of jet production, they can still constitute a major source of background in analyses involving electrons. In particular, a decay of $\pi^0 \to \gamma\gamma$, which happens to overlap with a charged particle track can readily fake the signature of an electron. In order to discriminate between electrons and jets, the following information may be used:

- **Shower shape** The fraction of the cluster energy in the EM layers, $\mathcal{F}_{\text{EM}}$, tends to be close to one for real electrons, and smaller for jets. The $\chi_{\text{EM}}^{2(7)}$ and $\chi_{\text{EM}}^{2(8)}$ variables are multidimensional discriminants that take into account both longitudinal and transverse shower shape information [5].
- **Calorimeter isolation** The calorimeter isolation for EM-clusters, $\mathcal{I}_{\text{cal}}^{\text{EM}}$, is defined as:

$$\mathcal{I}_{\text{cal}}^{\text{EM}} = \frac{E_T^{\text{tot}}(0.4) - E_T^{\text{EM}}(0.2)}{E_T^{\text{EM}}(0.2)},$$

where $E_T^{\text{tot}}(\Delta\mathcal{R})$ is the sum of transverse energies of all cells within in a cone of radius $\Delta\mathcal{R}$ around the EM cluster. $E_T^{\text{EM}}(\Delta\mathcal{R})$ restricts the sum to cells in the EM layers. Jets are expected to have a larger value of $\mathcal{I}_{\text{cal}}^{\text{EM}}$, than electrons from W and Z/γ^* decays.
- **Track isolation** The track isolation variable, $\mathcal{I}_{\text{trk}}^{\text{hc4}}$, is the p_T sum of all reconstructed tracks within a hollow cone of $0.05 < \Delta\mathcal{R} < 0.4$ around the candidate electron. Calculation of $\mathcal{I}_{\text{trk}}^{\text{hc4}}$ takes place at a stage when only tracks with $p_T > 0.5\,\text{GeV}$ are stored in the data. If the electron candidate has a central track match, then the isolation sum excludes tracks that do not originate from a common vertex.
- **$E_{\text{cal}}/p_{\text{trk}}$** Electrons which are matched to a central track are expected to deposit most of their energy in the EM layers. The momentum measured in the tracker is thus expected to be consistent with the energy measured in the calorimeter. Decays of $\pi^0 \to \gamma\gamma$ which overlap with random tracks are likely to have large values of $E_{\text{cal}}/p_{\text{trk}}$, since most tracks are at low p_T. Charged hadrons are likely to have low values of $E_{\text{cal}}/p_{\text{trk}}$ since they only deposit a small fraction of their energy in the EM layers of the calorimeter.
- **Multivariate classifiers** The $\mathcal{L}_{\text{elec}}^{(8)}$ variable is a likelihood based variable that combines tracking and shower shape information. Recently, a boosted decision tree (BDT) based variable was introduced to discriminate between electrons and hadronic jets [6]. Multivariate classifiers are mentioned again later in this chapter, and are used in the diboson analysis (see Chap. 7).

3.2.3 Muons

Firstly, track segments are reconstructed in each of the three muon layers from wire and scintillator hits. These track segments are combined to give a *local muon* candidate, which can be matched to a central track. An alternative algorithm starts from central track seeds, and looks for deposits of energy in the calorimeter that are consistent with the passage of a minimum ionising particle. *Loose*, *medium* and *tight* qualities have been defined for local muons based on the number of scintillator and wire hits [7]. In order to discriminate against background processes, the following variables are defined:

- **Calorimeter isolation** The calorimeter isolation, $\mathcal{I}_{cal}$ is defined as the E_T sum of all cells within a hollow cone of width $0.1 < \Delta\mathcal{R} < 0.4$ around the muon candidate.
- **Track isolation** The track isolation, $\mathcal{I}_{trk}$ is defined as the p_T sum of all tracks within a cone of width $\Delta\mathcal{R} < 0.5$ around the muon candidate. If the muon is matched to a central track, then this track is excluded from the isolation sum.

The following processes/decays are possible sources of background in analyses of Z/γ^* and W decays:

- **Heavy flavour decays** Heavy flavour hadrons (containing b or c quarks) can decay semileptonically to produce real muons. Since high p_T partons tend to produce hadronic jets, muons from semileptonic decays will tend to be surrounded by additional activity, and thus have larger values of $\mathcal{I}_{cal}$ and $\mathcal{I}_{trk}$.
- **In flight decays** Light charged mesons (e.g. $K^\pm$, $\pi^\pm$) can decay semileptonically within the tracking volume. The decay will tend to produce a "kink" in the track which can be identified by a relatively poor track fit χ^2. The discriminating power of the track fit χ^2 is significantly greater for tracks which have hits in the SMT.
- **Hadronic punch through** Charged hadrons (mostly p, $\pi^\pm$, $K^\pm$) can occasionally "punch" through the calorimeter, and thus make it to the muon system. These will be unlikely to pass through the toroid and thus be reconstructed in all layers of the muon system. Therefore, tighter local muon quality requirements may be imposed to reduce punch through contamination.
- **Cosmic ray muons** These are uncorrelated with the bunch crossing, and can be rejected based on timing information from the muon scintillator detectors, and by requiring that the track x and y coordinates are consistent with the beam position at that time of data taking (the average beam position for each run is measured during offline reconstruction).

3.2 Particle Reconstruction

3.2.4 Hadronic Taus

Tau leptons have a large branching fraction (roughly 60%) into hadrons. In D0, these decays are reconstructed with a simple calorimeter cone of size $\Delta\mathcal{R} = 0.3$. Tau candidates are classified into three categories based on their decay topology.

- **Type-1** A single track matched to a calorimeter cluster, but without an EM sub-cluster. These are mainly $\tau^\pm \to \pi^\pm \nu$ decays.
- **Type-2** As for type-1 except that there is at least one EM sub-cluster. These are mainly $\tau^\pm \to \rho^\pm \nu \to \pi^\pm \pi^0 \nu$ decays.
- **Type-3** Three reconstructed tracks within the cone. These are mainly $\tau^\pm \to \pi^\pm \pi^+ \pi^- \nu$ decays.

A neural network based classifier has been developed to separate hadronic taus from jets [8].

3.2.5 IC Electrons

The inter-cryostat region of the detector, $1.1 < |\eta_{det}| < 1.5$, has little or no EM calorimeter coverage. Electrons traversing this region are however likely to be reconstructed as hadronic taus. A recent development was to recover such "IC" electrons and to use a neural network discriminant, $NN_\tau(e)$, that is trained to separate hadronic jets and genuine IC electrons [9]. IC electrons that are based on type-2 taus (i.e., having an associated EM cluster) use the calorimeter energy for kinematic analysis. Those that are based on type-1 taus used the central track momentum. A comparison of the track and calorimeter energy resolutions for IC electrons can be found in Ref. [10]. IC electrons that are based on type-3 taus (i.e., more than one track within the reconstruction cone) are reclassified as type-1 or type-2.

3.2.6 Hadronic Jets

Hadronic jets are reconstructed as collimated deposits of energy in the calorimeter. Care must be taken in designing the algorithm used for reconstructing jets, such that their rates can be reliably calculated in QCD. A naive approach would be to simply sum up the energy of all cells within a cone of fixed size in $\Delta\mathcal{R}$ around a seed cell (or tower of cells) which is above a predefined threshold. Such an algorithm would not be considered "infra-red" safe, since the number of reconstructed jets is sensitive to the emission of very low energy particles. D0 uses a seed based mid-point cone algorithm [11] which is considered to be infra-red safe.

A key challenge in jet reconstruction is the energy scale calibration [12]. Dijet, γ+jet and Z/γ^*+jet events are all used in order to determine the energy scale which corrects back to the particle level jets. This calibration relies on having used the $Z/\gamma^* \to e^+ e^-$ lineshape to calibrate the EM energy scale.

3.2.7 Missing Transverse Energy

The missing transverse energy, $\slashed{E}_T$, is first reconstructed by summing vectorially the E_T of each calorimeter cell. This requires knowledge of the event primary vertex in order to translate from a cell energy to a cell E_T. Different layers of the calorimeter can be included/excluded depending on the application. Typically, the coarse hadronic layers are excluded since they tend to contribute significant additional noise. Additional calibrations and corrections to jets, photons, electrons and taus can be propagated to the $\slashed{E}_T$. High p_T muons deposit only a small amount of energy in the calorimeter, typically a few GeV. This energy (estimated on average from simulation) is replaced by the momentum measured by the inner tracker.

3.3 Monte Carlo Simulation

Most analyses rely heavily on simulations of physics processes and the detector response, based on Monte Carlo (MC) event generators as introduced in Chap. 1. For example, cross section measurements rely on simulations to predict background rates, and to determine corrections for signal acceptance and efficiency.

An example event generator used by DØ to generate many different physics processes is PYTHIA (see Chap. 1). The stable[2] produced particles are fed through a GEANT [14] based simulation of the detector response. The resulting output is then *digitised* into the same format as the real data. The digitisation stage also merges MC events with randomly triggered bunch crossings from real data, to simulate the effects of additional $p\bar{p}$ collisions in the same bunch crossing (called "pile-up"). The digitised output can be reconstructed and analysed using the same software as used for the real data.

Unfortunately, the simulation is found to be rather inaccurate in a number of areas. For example, the p_T resolution of muons and energy resolution of electrons are over-optimistic. Reconstruction and quality cut efficiencies for electrons and muons are typically overestimated by the simulation. Corrections for the above inaccuracies need to be applied, and tend to inflate the systematic uncertainties in physics analyses. Furthermore, there are known flaws in the physics modellings of certain processes. For example the D0 implementation of PYTHIA[3] provides an inaccurate description of the p_T distribution of W and Z/γ^* events. In order to improve the description, weights are applied to the simulated events. This "re-weighting" is typically based on comparison with data or with more accurate MC predictions, e.g. from the RESBOS program [16].

---

[2] A particle is typically considered stable if its lives long enough to be detectable, e.g. $c\tau > 10$ mm [13].

[3] D0 uses version v6_413 and tune A [15].

3.4 Unfolding

Measurement of a differential cross section inevitably requires correction of the observed distribution for the effects of detector resolution and efficiency. For example, a steeply falling distribution is "smeared" by detector resolution. One may also correct for experimental acceptance, though this is generally not a good idea, since it requires additional theoretical input to extrapolate from the observable phase space to, e.g. 4π acceptance.

A simple method would be to simulate the physics process (e.g. $Z/\gamma^* \to e^+e^-$) with a Monte Carlo event generator, and produce a "generator-level" distribution (e.g. of the dilepton p_T). The same events can be fed into a simulation of the detector response and selection efficiency, giving a second "detector-level" distribution. The ratio of these two histograms provides a simple "bin-by-bin" correction for resolution and efficiency. Unfortunately, this method is intrinsically biased. If the bin-to-bin migration due to resolution is significant, then the correction factor for each bin depends on the shape of the generator-level distribution.

A commonly used technique is to generate a 2D migration matrix from a simulation of the detector response. This matrix can be inverted, and used to unfold the data [17]. The inversion introduces numerical instabilities, and various techniques have been introduced to control these, for example the singular value decomposition (SVD) approach as implemented in the GURU code [18]. These methods inevitably inflate (and also correlate) the bin-by-bin statistical uncertainties. If the effect of bin-to-bin migration is minimal, then the simple bin-by-bin unfolding is sufficient. This is one of the motivations for introducing new variables for studying the Drell-Yan p_T (see Chap. 5).

3.5 Multivariate Classifiers

Multivariate classifier algorithms use information from many different variables, including their correlations, to produce a single variable with optimal discrimination between signal and background processes. Many of the currently available algorithms have been implemented in the TMVA package for use in high energy physics [19]. The diboson analysis of Chap. 7 makes extensive use of this package.

References

1. K. Peters, J. Kvita, D0 Note 5005, D0 (2006)
2. H. Yin, Measurement of the Forward-Backward Charge Asymmetry using $p\bar{p}$ to Z/γ^* to e^+e^- events at $\sqrt{s} = 1.96$ TeV, PhD thesis, University of Science and Technology of China (2010)
3. K. Nakamura et al., J. Phys. G **37**, 075021 (2010)

4. R. Wigmans, *Calorimetery, Energy Measurement in Particle Physics* (Oxford Science Publications, Oxford, 2000)
5. A. Abdessalam, D0 Note 3745, D0 (2000)
6. T. Head, X. Bu, K.A. Petridis. A Multivariate based Electron ID, D0 Note 6238, D0 (2011)
7. P. Calfayan and others, D0 Note 5157, D0 (2007)
8. S. Protopopescu, P. Svoisky, D0 Note 5094, D0 (2006)
9. B. Calpas, J. Kraus, T. Yasuda, D0 Note 6051, D0 (2010)
10. M. Vesterinen, D0 Note 6018, D0 (2011)
11. G.C. Blazey et al., in *Proceedings of the Workshop QCD and Weak Boson Physics in Run II*, (2000), p. 47
12. V.M. Abazov et al., Phys. Rev. Lett. **101**, 062001 (2008)
13. J.M. Butterworth et al., The Tools and Monte Carlo Working Group Summary Report from the Les Houches 2009 Workshop on TeV Colliders. In Les Houches 2009 Tools and Monte Carlo working group, arXiv:1003.1643v1 [hep-ph] 2010
14. S. Agostinelli et al., Nucl. Inst. Meth. in Phys. Res. A **506**, 250–303 (2003)
15. R. Field and R. Craig Group. Pythia tune A, Herwig, and Jimmy in Run 2 at CDF. CDF-ANAL-CDF-PUBLIC 7822, CDF (2005)
16. C. Balazs, C-P. Yuan, Phys. Rev. D **56**, 5558–5583 (1997). We use the CP version of the code and grid files
17. G. D'Agostini, A multidimensional unfolding method based on Bayes' theorem. Nucl. Instrum. Meth. A **362**, 487–498 (1995)
18. A. Hocker, V. Kartvelishvili, Svd approach to data unfolding. Nucl. Instrum. Meth. A **372**, 469–481 (1996)
19. A. Hoecker, P. Speckmayer, J. Stelzer, J. Therhaag, E. von Toerne, H. Voss. CERN-OPEN 007 (2007)

Chapter 4
Electron and Photon Energy Calibration

4.1 Introduction

As mentioned in Chap. 2, the EM layers of the central calorimeter (CC) are divided into 32 azimuthal modules, and the energy response is degraded near to the module boundaries, or ϕ-gaps. In addition to the degraded energy response, the centroid of EM clusters is biased away from gaps. The ϕ_{mod} variable is defined as:

$$\phi_{\text{mod}} = \left| \text{mod}(32\phi/2\pi, 1) - \frac{1}{2} \right|,$$

such that the boundaries lie at $\phi_{\text{mod}} = 0.5$. The regions, $\phi_{\text{mod}} > 0.4$ have traditionally been referred to as "non-fiducial". Most analyses of final states involving electrons and photons have so far excluded these regions, thus sacrificing valuable acceptance.

Chapter 5 introduces the ϕ_η^* variable as being optimal from an experimental point of view for studying the Drell-Yan p_T distribution. Whilst ϕ_η^* is relatively insensitive to the effects of lepton momentum resolution and isolation efficiencies, it is particularly sensitive to gaps in the detector that are back-to-back in ϕ. As we shall see later, such gaps cause the event selection efficiency to modulate as a function of ϕ_η^*. Improving the treatment of the CC ϕ-gaps therefore increases the achievable precision of the ϕ_η^* measurement.

Figure 4.1 shows the dielectron invariant mass (M_{ee}) distribution in a sample of $Z/\gamma^* \to e^+e^-$ data events that contain two CC electrons. The distribution is shown separately for events in which 0, 1, and 2 electrons are in a ϕ-gap. Clearly, electrons in the ϕ-gaps suffer significant energy losses. Since the $Z/\gamma^* \to e^+e^-$ physics can be assumed to be ϕ-independent (due to the unpolarised beams), the observed $Z/\gamma^* \to e^+e^-$ lineshape can be used to correct for these energy losses. It should be noted that a correction is already implemented as part of the reconstruction. Figure 4.2 shows the correction factor applied to EM cluster energies as a function of ϕ_{mod}. This correction was determined early in Run II with rather low statistics, and clearly under corrects as can be seen in Fig. 4.1. In addition, the non-

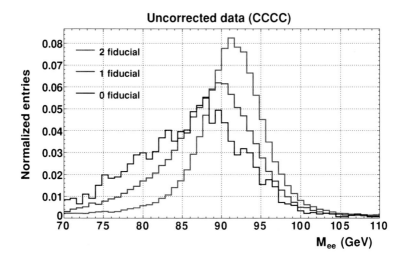

Fig. 4.1 The M_{ee} distribution in a sample of $Z/\gamma^* \to e^+e^-$ events in data, where both electrons are in the CC, and separating events with 0,1 and 2 fiducial electrons. All three distributions are normalised to unit area

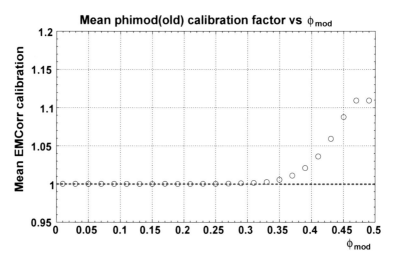

Fig. 4.2 The mean correction factor of ϕ-gap energy losses, which is currently applied as part of the reconstruction, but under corrects

Gaussian tails in the resolution are poorly modelled by the MC, partly because the finite charge collection time is ignored.

The aim of this study is to:

- Correct for the ϕ-bias of the EM clusters.
- Correct for the energy losses in the ϕ-gaps.
- Improve the description of the energy resolution in the MC.

## 4.1 Introduction

This will hopefully allow many analyses to increase their electron/photon acceptance by relaxing the fiducial requirements, thus improving their sensitivity. Those analyses that already include these regions will benefit from improved energy resolution and more accurate modelling by the MC.

4.2 Dataset

Run numbers from 151817 to 252918 are included (19th April 2002–13th June 2009). The data taking period up to March 2006 is referred to as Run IIa, and the period after as Run IIb. Certain periods of data taking are flagged as bad by the D0 data quality group, and are removed. In addition, a large sample of $Z/\gamma^* \rightarrow e^+e^-$ events is generated with PYTHIA within the standard D0 MC framework.

4.3 Correction of the ϕ-Bias

For track matched electrons, the direction of the electron is determined from the track, which has significantly better angular resolution than the cluster. However, for photons and electrons without track matches, the particle direction must be determined from the centroid of the cluster and the position of a primary interaction vertex. In the CC, the calorimeter cluster ϕ is biased away from the module boundaries. For the purposes of correcting for the energy losses in the ϕ-gaps, we need a consistent definition of the ϕ_{mod} variable for all clusters.[1] Central tracks can be extrapolated to the EM3 layer of the calorimeter (the layer with the finest granularity) as follows:

$$\tan \phi_{trkEM3} = \frac{R_{CC} \times \sin \phi' - Y_{shift}^{CC}}{R_{CC} \times \cos \phi' - X_{shift}^{CC}},$$

where

$$\phi' = \phi_{trk} + \frac{AA \times E_T}{R_{CC} \times q \times B_z},$$

where E_T is the transverse energy of the EM cluster, q is the charge of the track, and B_z is the sign of the solenoid polarity. The following parameters are provided by a group working on a measurement of m_W [1]:

- $AA = 21.0$ cm.
- $R_{CC} = 91.9$ cm,
- $Y_{shift}^{CC} = -0.46$ cm,
- $X_{shift}^{CC} = -0.33$ cm.

---

[1] In this chapter, and the two physics analysis presented later in this thesis, all electrons are required to have track matches. However, the corrections described here are intended for use in all analyses involving EM clusters.

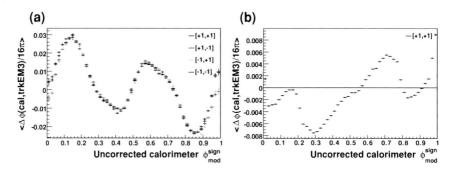

Fig. 4.3 The mean ϕ-shift, $32(\phi_{trkEM3} - \phi_{cal})/2\pi$ as a function of ϕ_{mod}^{sign} in (*left*) data and *right* $Z/\gamma^* \to e^+e^-$ MC. The legend entries indicate the values of B_z (1st element) and q (2nd element). **a** Data **b** MC

We are now interested in the sign of the ϕ_{mod} variable and define,

$$\phi_{mod}^{sign} = \text{mod}(32\phi_{trkEM3}/2\pi, 1).$$

Figure 4.3 shows separately for data and MC, the mean ϕ-shift, $32(\phi_{trkEM3} - \phi_{cal})/2\pi$, as a function of the uncorrected calorimeter ϕ_{mod}^{sign}, of track matched electrons in a sample of $Z/\gamma^* \to e^+e^-$ events. The magnitude of the ϕ-shift is a factor of ~ 4 times larger in data, than in MC. We can use the mean ϕ-shift as a function of ϕ_{mod}^{sign} to correct the cluster position.

For track matched clusters, ϕ_{mod} is now determined from the track, extrapolated to the EM3 layer of the calorimeter. For non track matched clusters, ϕ_{mod} is determined from the corrected cluster position.

4.4 Energy Calibration

4.4.1 Additional Calibrations

In addition to correcting for the ϕ_{mod} dependence of the energy scale, three further corrections are implemented:

- Alternative cell-by-cell gain constants, and energy loss corrections from a group working on a W boson mass and width measurement [1]. The original energy loss corrections are severely overestimated due to inaccuracies in the simulated detector response. Whilst this is mostly absorbed by a corresponding reduction of the gain calibration factors, the energy response suffers from non uniformities that cannot be tolerated for the m_W measurement.

4.4 Energy Calibration

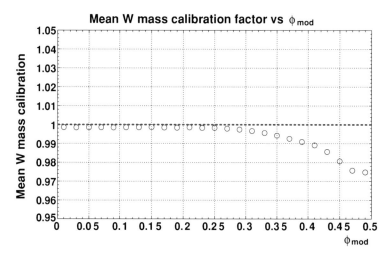

Fig. 4.4 Mean W mass group calibration factor as a function of ϕ_{mod}

- An instantaneous luminosity dependent calibration. The MC predicts an increase in the average EM energy with $\mathcal{L}_{inst}$, as expected due to the extra energy from pile-up. However, the data displays the opposite trend. An average correction for this dependence is determined using the E_{cal}/p_{trk} in $W \to e\nu$ events and is documented in Ref. [2].
- Individual cells in the calorimeter are known to saturate at roughly 200 GeV, but this is not modelled in the simulation. A simple correction truncates the energy of any cells in the MC that exceed the saturation value for that cell.

Interestingly, the m_W group correction factors introduce a ϕ_{mod} dependence of up to $\approx 3\%$, as shown in Fig. 4.4. This can be understood as follows: the reduced energy loss corrections have a larger effect on ϕ-gap clusters since they tend to be reconstructed with lower energies on average.

4.4.2 A ϕ_{mod} Dependent Energy Correction

An energy scale correction for both data and MC, as a function of ϕ_{mod} is determined using $Z/\gamma^* \to e^+e^-$ events as follows:

- A clean sample of $Z/\gamma^* \to e^+e^-$ events is obtained by requiring two high E_T ($E_T > 20$ GeV) electrons, in the CC or EC, both of which are matched to central tracks. One electron is required to be within the CC region.
- Events are required to have fired any one of the single EM triggers (not for the MC).

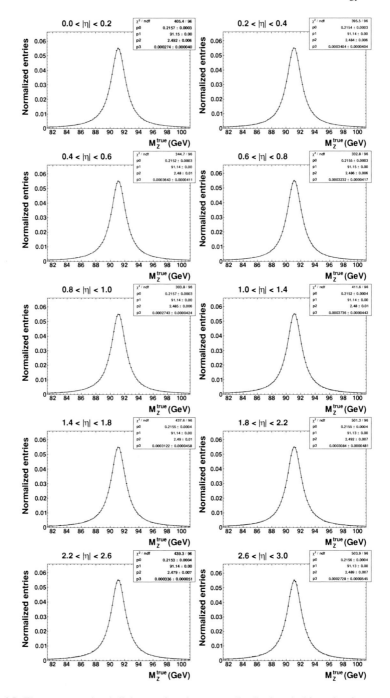

Fig. 4.5 The generator level dielectron invariant mass distribution, in bins of η *for one of the electrons*. The distributions are fit with a Breit-Wigner on top of a linear function

4.4 Energy Calibration

- Both electron candidates must satisfy the following shower shape and isolation requirements (see Chap. 3 for details on the selection variables):
  - $\mathcal{F}_{EM} > 0.9$,
  - $\mathcal{I}_{cal}^{EM} < 0.1$,
  - $\mathcal{I}_{trk}^{hc4} < 2\,\text{GeV}$.
- One "tag" electron is required to be in EC or in CC fiducial ($\phi_{mod} < 0.4$), and satisfy $\chi_{EM}^{2(7)} < 12$.
- The other "probe" electron is required to be in CC. Note that this procedure has nothing to do with "tag and probe" measurements of efficiencies (see e.g., Chap. 6).
- Both electrons are considered as candidate tags and probes.
- As a function of probe electron ϕ_{mod}, the $Z/\gamma^* \to e^+e^-$ peak position, M_Z, is extracted in a fit to the lineshape.
- The resulting energy scale correction to CC electrons is $(m_Z/M_Z(\phi_{mod}))^2$, where m_Z and M_Z are the expected and observed $Z/\gamma^* \to e^+e^-$ peak positions respectively.

The power of 2 in the correction is motivated as follows: the dielectron invariant mass can be written as

$$M_{ee} = \sqrt{2E_1 E_2(1 - \cos(\Delta\theta))},$$

where $\Delta\theta$ is the opening angle between the two electrons. If a ϕ_{mod} dependent scale-factor, $f(\phi_{mod})$ is applied to *only one electron* then,

$$M'_{ee}(\phi_{mod}) = \sqrt{2E_1 f(\phi_{mod}) E_2(1 - \cos(\Delta\theta))}.$$

$f(\phi_{mod})$ can therefore be determined from

$$\frac{M'_{ee}(\phi_{mod})}{M_{ee}} \sim \sqrt{f(\phi_{mod})}.$$

However, the power of 2 in the correction is only an approximation, since it is based on the assumption that the energy scale correction is to be applied to only one of the two electrons in the event and, in fact, the ϕ_{mod} of the two electrons are highly correlated. The idea of this first energy scale correction is to parameterise the largest part of the ϕ_{mod} dependence, and in addition the shower shape dependence. Residual ϕ_{mod} dependence is expected, due to these correlations, and will be later removed using an iterative method.

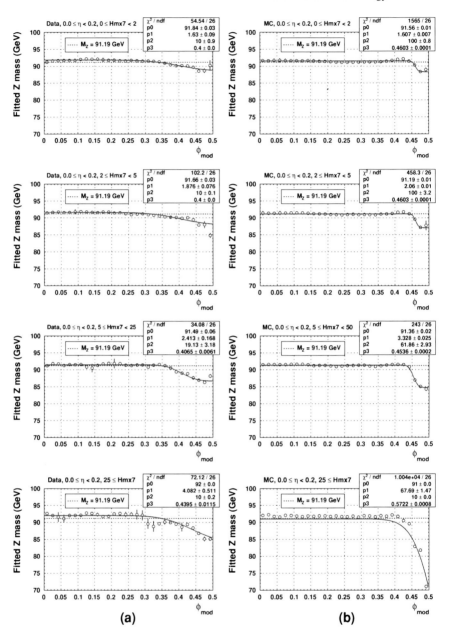

Fig. 4.6 Fitted Z peak position versus ϕ_{mod} for the region $0 < \eta < 0.2$. The rows correspond to different ranges of $\chi^{2(7)}_{\text{EM}}$. **a** Data **b** MC

4.4 Energy Calibration

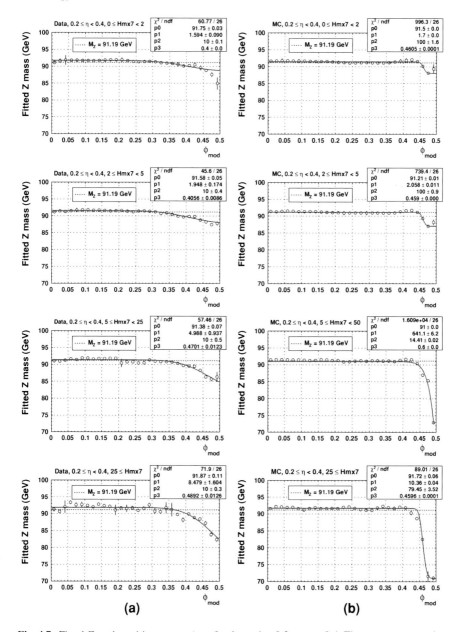

Fig. 4.7 Fitted Z peak position versus ϕ_{mod} for the region $0.2 < \eta < 0.4$. The rows correspond to different ranges of $\chi^{2(7)}_{\text{EM}}$. **a** Data **b** MC

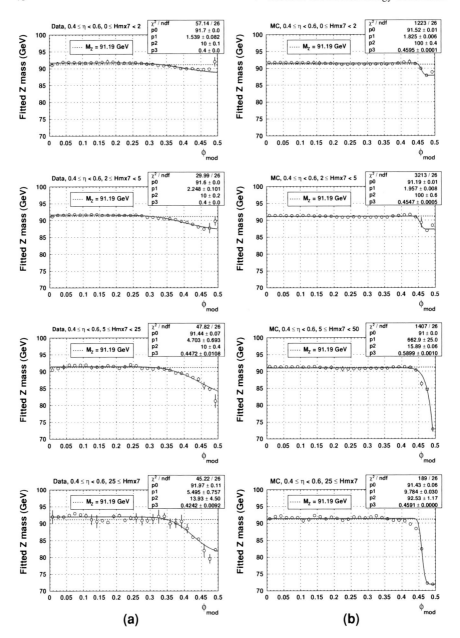

Fig. 4.8 Fitted Z peak position versus ϕ_{mod} for the region $0.4 < \eta < 0.6$. The rows correspond to different ranges of $\chi^{2(7)}_{\text{EM}}$. **a** Data **b** MC

4.4 Energy Calibration

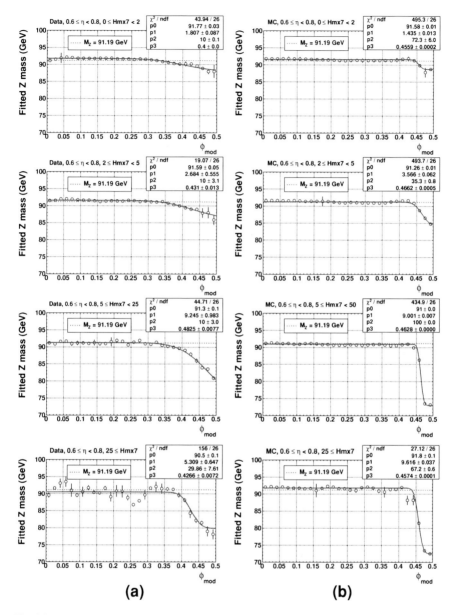

Fig. 4.9 Fitted Z peak position versus ϕ_{mod} for the region $0.6 < \eta < 0.8$. The rows correspond to different ranges of $\chi^{2(7)}_{\text{EM}}$. **a** Data **b** MC

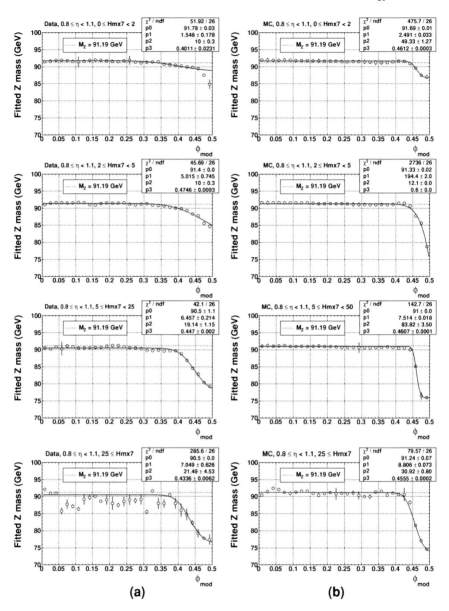

Fig. 4.10 Fitted Z peak position versus ϕ_{mod} for the region $\eta > 0.8$. The rows correspond to different ranges of $\chi^{2(7)}_{EM}$. **a** Data **b** MC

4.4.3 Shower Shape Dependence

The EM cluster shower shape may provide additional information to estimate the amount of energy loss in the ϕ-gaps. We consider the dependence on the $\chi^{2(7)}_{\mathrm{EM}}$ variable (see Chap. 3). The background from multijet events is still expected to be small when the probe electron has large $\chi^{2(7)}_{\mathrm{EM}}$, given the tight quality requirements on the tag electron.

4.4.4 Pseudorapidity Dependence

Additional η dependence may be expected, since at larger incident angles, particles traverse more material before entering the calorimeter, thus changing the shower profile. Before studying the η dependence, as determined at the reconstructed EM object level, it is important to understand any possible dependence of the physics. Figure 4.5, shows the generator level M_{ee} distributions, as predicted by PYTHIA, separated into bins of η of *one electron*, where the other electron has no η constraint (i.e., two entries per event). The distributions are fitted with a Breit-Wigner on top of a linear function, and any dependence of the Breit-Wigner mass (parameter p_1 in Fig. 4.5) is found to be less than 0.01 %. Correcting for additional η dependence is therefore unlikely to introduce any significant model dependence.

4.4.5 Fits for the ϕ_{mod} Dependence

Figures 4.6, 4.7, 4.8, 4.9 and 4.10 show the ϕ_{mod} dependence of the fitted M_Z, in bins of $\chi^{2(7)}_{\mathrm{EM}}$ and η. The left hand columns show the data and the right hand columns show the MC. The following fit function is used:

$$f(\phi_{\mathrm{mod}}) = p_0 + p_1 e^{(p_2 \phi_{\mathrm{mod}} - p_3)}.$$

The fit values are displayed on the top right hand corners of the plots.

4.4.6 Iteration of the Corrections

As mentioned earlier, the ϕ correlations between the tag and probe electrons may leave some residual ϕ_{mod} dependence of the energy scale after correction. After applying the ϕ_{mod} and $\chi^{2(7)}_{\mathrm{EM}}$ dependent scale correction, a second binned correction (in ϕ_{mod} only) is determined using events in which both electrons are in the CC region. This part of the correction is determined separately for Run IIa and Run IIb data/MC. The correction is determined in the same way except that it is applied as $m_Z/M_Z(\phi_{\mathrm{mod}})$, instead of $m_Z^2/M_Z^2(\phi_{\mathrm{mod}})$. In addition, the fiducial requirement for the tag electron is relaxed, since the dependence of the energy scale on ϕ_{mod} is mostly

50   4 Electron and Photon Energy Calibration

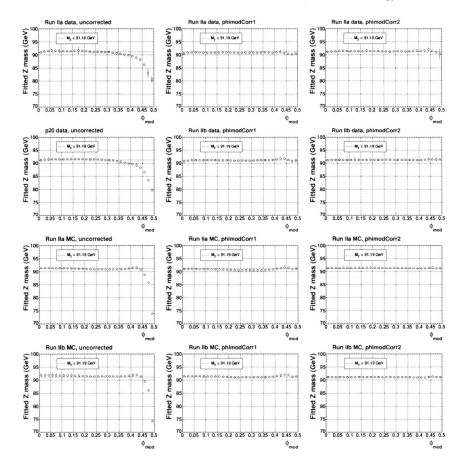

Fig. 4.11 The ϕ_{mod} dependence of fitted M_Z (*left*) before calibration, (*middle*) after the first parameterised calibration, and (*right*) after the final binned correction. The four rows correspond to Run IIa data, Run IIb data, Run IIa MC, and Run IIb MC

removed by the first correction. Figure 4.11 shows a small residual dependence of the energy scale on ϕ_{mod} after the two dimensional (ϕ_{mod}, $\chi_{EM}^{2(7)}$) correction. After only one iteration of the one dimensional binned correction in ϕ_{mod}, the maximum variations with ϕ_{mod} are within 0.3%, which is considered to be adequate.

4.4.7 Resolution Improvements

Figure 4.12 shows for (left hand column) data and (right hand column) MC, the invariant mass distributions of CCCC[2] events before and after correction. The top

---

[2] We refer to events with two CC electrons as CCCC; those with one CC and one EC electron as CCEC; and those with two EC electrons as ECEC.

4.4 Energy Calibration

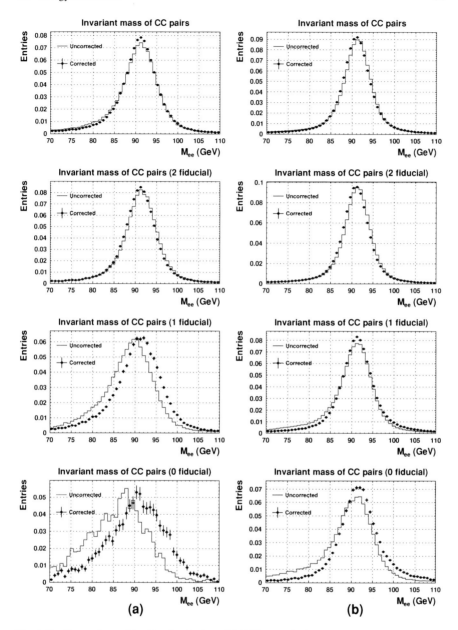

Fig. 4.12 Comparison of the M_{ee} distribution of CCCC events before and after calibration in data. The rows require different numbers of fiducial cluster as indicated in the figure titles (the *top row* is inclusive). **a** Data **b** MC

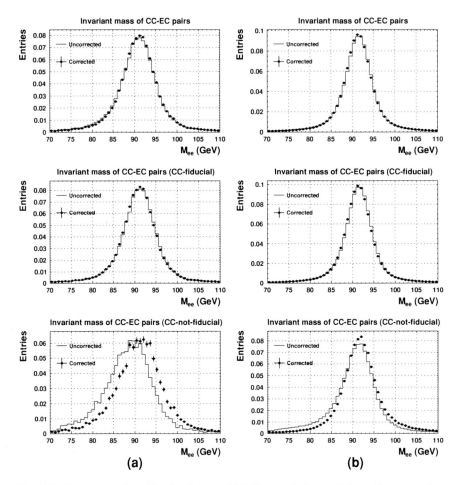

Fig. 4.13 Comparison of the M_{ee} distribution of CCEC events before and after calibration in data. The rows require different numbers of fiducial cluster as indicated in the figure titles (the *top row* is inclusive). **a** Data **b** MC

row includes all CC electrons, and shows a significant improvement in the height of the Z peak, and reduction in the tail on the left hand side of the peak. The second, third, and fourth rows divide the sample into 0, 1, and 2 fiducial electrons. Figure 4.13 shows the equivalent distributions for CCEC events. Figure 4.14 shows the CCCC lineshape before and after correction, with a fit based on a Breit-Wigner convoluted with a Crystal-Ball function [3]. The Crystal-Ball function is discussed extensively in the context of applying energy smearing to the MC simulation (see Sect. 4.5).

4.4 Energy Calibration

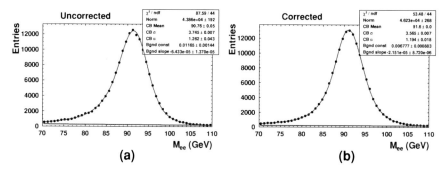

Fig. 4.14 Invariant mass distributions of CCCC dielectron events (*left*) before calibration and (*right*) after calibration. **a** Before correction **b** after correction

4.4.8 Energy Dependence

It is important to check that despite resolution improvements at the m_Z scale, the resolution at higher energies is not degraded. This would of course reduce the sensitivity of searches for heavy dielectron and diphoton resonances. Figures 4.15 and 4.16 compare the E_T distributions of fiducial and non-fiducial CC electrons, in bins of M_{ee}, before and after the calibration. Before calibration, the distribution is softer for non-fiducial electrons in all M_{ee} bins. After calibration, the agreement is improved not only in the bin from 80 to 100 GeV, which roughly corresponds to where the calibration was determined, but also in all other mass bins.

4.5 Monte Carlo Over-Smearing

As discussed earlier, the energy resolution predicted by the GEANT based simulation of the D0 detector, is significantly better than that observed in data. It is therefore necessary to apply additional, so called "over-"smearing, on top of the simulated energy resolution. The over-smearing is currently applied as [4]:

$$\frac{E'}{E} = 1 + x_{\text{Gaus}}(0, C),$$

where $x_{\text{Gaus}}(a, b)$ is a random number drawn from a Gaussian distribution with mean a and width b. In addition, an energy scale correction,

$$E' = \alpha E,$$

is applied before the smearing. This form of the over-smearing does not accurately reproduce the non-Gaussian tails observed in the data, especially in the CC ϕ-gaps.

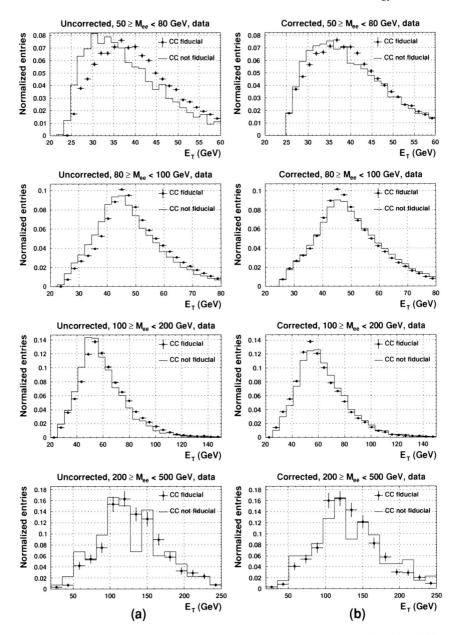

Fig. 4.15 The E_T distribution before and after calibration in data, and separated into bins of M_{ee}. **a** Before correction **b** after correction

4.5 Monte Carlo Over-Smearing

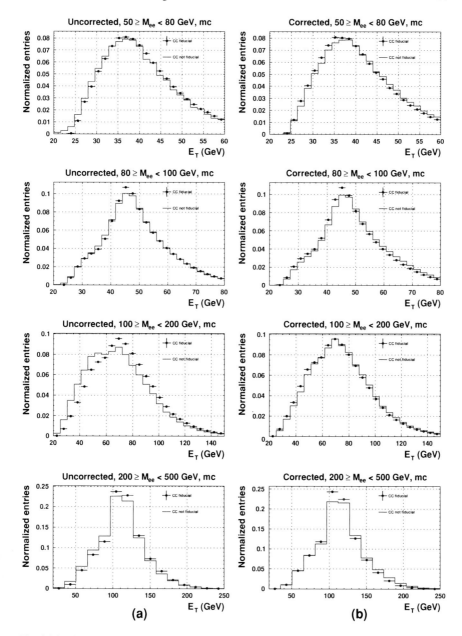

Fig. 4.16 The E_T distribution before and after calibration in MC, and separated into bins of M_{ee}. **a** Before correction **b** after correction

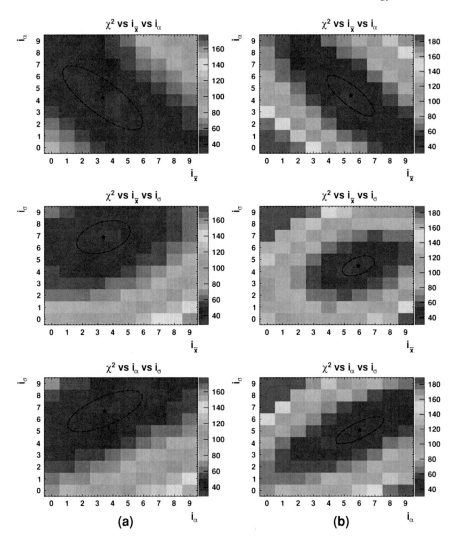

Fig. 4.17 Two dimensional projections of the χ^2 space for each pair of the CCfid smearing parameters, with the third parameter fixed at the value giving the minimum χ^2. The star, and ellipse represent the best fit and 68 % C.L. contours respectively. **a** Run IIa **b** run IIb

4.5.1 The Crystal Ball Function

As discussed earlier, the resolution also contains non-Gaussian "lossy" tails due to incomplete charge collection. A more general function with a Gaussian core portion, and a power law tail was introduced by the Crystal Ball experiment at SLAC [3]:

4.5 Monte Carlo Over-Smearing

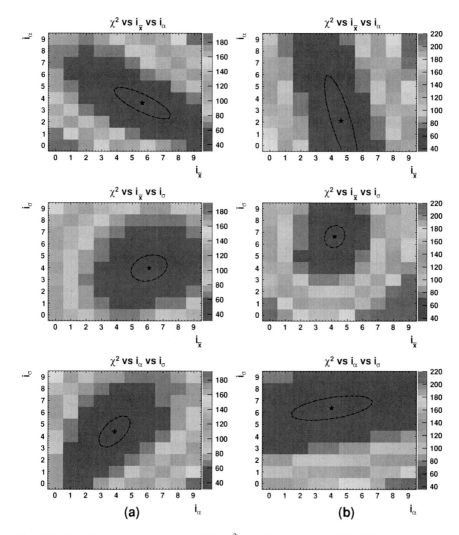

Fig. 4.18 Two dimensional projections of the χ^2 space for each pair of the CCnonfid smearing parameters, with the third parameter fixed at the value giving the minimum χ^2. The star, and ellipse represent the best fit and 68 % C.L. contours respectively. **a** Run IIa **b** run IIb

$$f(x; \alpha, n, \bar{x}, \sigma) = \begin{cases} \exp\left(-\dfrac{(x-\bar{x})^2}{2\sigma^2}\right), & \text{for } \dfrac{x-\bar{x}}{\sigma} > -\alpha \\ \left(\dfrac{n}{|\alpha|}\right)^n \exp\left(-\dfrac{\alpha^2}{2}\right)\left(\dfrac{n}{|\alpha|} - |\alpha| - \dfrac{x-\bar{x}}{\sigma}\right)^{-n}, & \text{for } \dfrac{x-\bar{x}}{\sigma} \leq -\alpha \end{cases}$$
(4.1)

- The σ parameter determines the width of the Gaussian core part of the resolution.
- The α parameter controls the width of the power law lossy tail of the resolution.

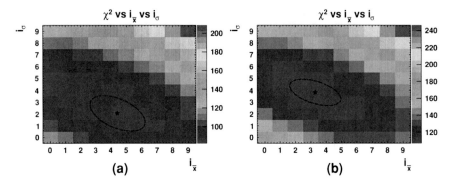

Fig. 4.19 Two dimensional projections of the χ^2 space for each pair of the EC smearing parameters, with the third parameter fixed at the value giving the minimum χ^2. The star, and ellipse represent the best fit and 68% C.L. contours respectively. **a** Run IIa **b** run IIb

Table 4.1 Fit ranges, values and uncertainties for the smearing parameters

Category	Param.	Start	Step	Value	Uncertainty
		Run IIa			
CCfid	$\bar{x}$	1.0057	0.0003	1.007	0.001
CCfid	σ	0.0126	0.0006	0.017	0.002
CCfid	α	0.96	0.02	1.03	0.09
CCnonfid	$\bar{x}$	1.011	0.002	1.023	0.005
CCnonfid	σ	0.030	0.003	0.043	0.008
CCnonfid	α	0.7	0.07	0.83	0.20
EC	$\bar{x}$	0.991	0.0007	0.989	0.001
EC	σ	0.0220	0.0013	0.025	0.002
		Run IIb			
CCfid	$\bar{x}$	1.0076	0.0003	1.0100	0.0007
CCfid	σ	0.011	0.0006	0.013	0.001
CCfid	α	0.88	0.02	0.91	0.05
CCnonfid	$\bar{x}$	1.013	0.001	1.021	0.003
CCnonfid	σ	0.033	0.002	0.037	0.005
CCnonfid	α	0.6	0.001	0.74	0.07
EC	$\bar{x}$	0.9890	0.0007	0.9890	0.0007
EC	σ	0.0220	0.0013	0.023	0.001

- The n parameter governs the shape of the power law lossy tail. Smaller values of n correspond to a "longer" tail.
- The $\bar{x}$ parameter is the mean of the Gaussian core part of the resolution. Typically, an increase in the width of the lossy tail, needs to be compensated by an increase in the mean.

4.5 Monte Carlo Over-Smearing

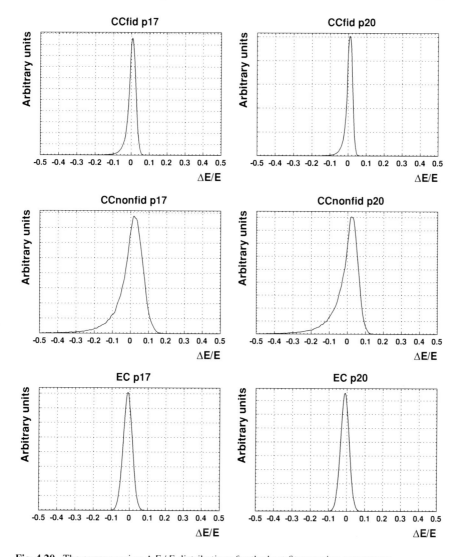

Fig. 4.20 The oversmearing $\Delta E/E$ distributions for the best fit smearing parameters

4.5.2 Method to Fit for the Crystal Ball Parameters

The fit is performed by varying the Crystal Ball parameters applied to the MC, and minimising the χ^2 between the data and MC, in the M_{ee} distribution around m_Z. Since there is enough freedom in the other three parameters, to adequately describe the data, the n parameter is fixed at $n = 7$, reducing the number of free parameters to three. This value of n is found to give reasonable results, and further variation is considered to be beyond the scope of this study. A $10 \times 10 \times 10$ grid $(\bar{x}, \sigma, \alpha)$ of MC

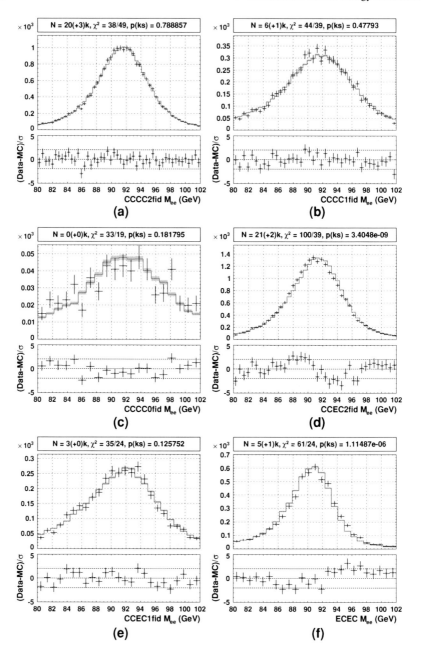

Fig. 4.21 Data-vs-MC comparison of the dielectron mass distribution using the best fit smearing parameters for MC (Run IIa data and MC). **a** CCfid+CCfid **b** CCfid+CCnonfid **c** CCnonfid+CCnonfid **d** CCfid+EC **e** CCnonfid+EC **f** EC+EC

4.5 Monte Carlo Over-Smearing

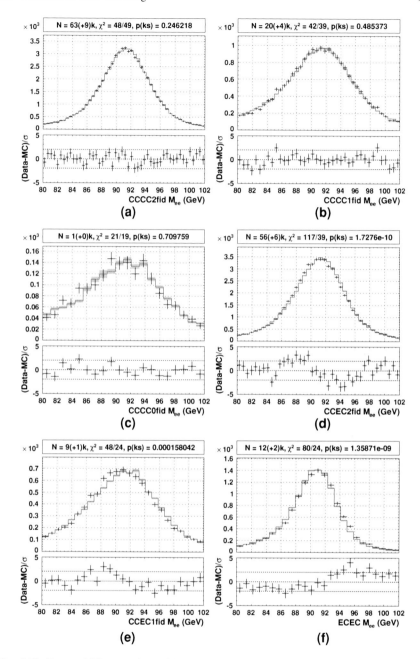

Fig. 4.22 Data-vs-MC comparison of the dielectron mass distribution using the best fit smearing parameters for MC (Run IIb data and MC). **a** CCfid+CCfid **b** CCfid+CCnonfid **c** CCnonfid+CCnonfid **d** CCfid+EC **e** CCnonfid+EC **f** EC+EC

M_{ee} histograms is generated, from which a grid of data-vs-MC χ^2 can be calculated. The χ^2 space is then fitted with the following function:

$$f(x, y, z) = \chi^2_{\min} + [(x - x_0)/\sigma_x]^2 + [(y - y_0)/\sigma_y]^2 + [(z - z_0)/\sigma_z]^2 \\ - 2\rho_{xy}(x - x_0)(y - y0)/(\sigma_x \sigma_y) \\ - 2\rho_{xz}(x - x_0)(z - z0)/(\sigma_x \sigma_z) \\ - 2\rho_{yz}(y - y_0)(z - z0)/(\sigma_y \sigma_z)$$

The parameters are fitted separately for Run IIa and Run IIb, and for the following three categories of EM cluster:

- **Category 1: CCfid** CC-fiducial clusters are defined as; $|\eta_{det}| < 1.1$ and $\phi_{mod} < 0.4$. The parameters are fitted using events in which both electrons are CC-fiducial.
- **Category 2: CCnonfid** CC-non-fiducial clusters are defined as; $|\eta_{det}| < 1.1$ and $\phi_{mod} \geq 0.4$. The parameters are fitted using events containing two CC electrons, where *at least one* is non-fiducial. Any CCfid electrons are smeared using their already tuned parameters.
- **Category 3: EC** EC clusters are defined as having $|\eta_{det}| > 1.5$. The parameters are fitted using events containing CC-fiducial, and EC electrons, in which *one or both* electron may be EC. For EC clusters, a simple Gaussian smearing is used, and the fit is only in two dimensions $(\bar{x}, \sigma)$.

4.5.3 Results

Figures 4.17 and 4.18 show, for the CCfid and CCnonfid categories respectively, the three different 2D projections of the data-vs-MC χ^2 having fixed the other parameter to its best fit value. The ellipse indicates the 68 % C.L. interval. Figure 4.19 shows the data-vs-MC χ^2 as a function of the two fit parameters for EC electrons.

Table 4.1 lists the best fit values of the smearing parameters, and their 68 % C.L. uncertainties. Also provided are the parameter start values and step-sizes used in the fits. Figure 4.20 shows the (over-smearing) $\Delta E/E$ distributions for the best fit parameters. The correlation matrices $(\bar{x}, \sigma, \alpha)$ for CCfid and CCnonfid clusters are:

Run IIa CCfid:
$$\begin{pmatrix} 1 & & \\ 0.428 & 1 & \\ -0.843 & 0.759 & 1 \end{pmatrix}$$

Run IIb CCfid:
$$\begin{pmatrix} 1 & & \\ 0.466 & 1 & \\ -0.890 & 0.840 & 1 \end{pmatrix}$$

Run IIa CCnonfid:
$$\begin{pmatrix} 1 & & \\ 0.229 & 1 & \\ -0.860 & 0.580 & 1 \end{pmatrix}$$

Run IIb CCnonfid:
$$\begin{pmatrix} 1 & & \\ 0.121 & 1 & \\ -0.726 & 0.566 & 1 \end{pmatrix}$$

4.5 Monte Carlo Over-Smearing

And the correlation matrices $(\bar{x}, \sigma)$ for EC clusters are:

$$\text{Run IIa:} \quad \begin{pmatrix} 1 & \\ 0.049 & 1 \end{pmatrix} \quad \text{Run IIb:} \quad \begin{pmatrix} 1 & \\ 0.017 & 1 \end{pmatrix}$$

4.5.4 Data Versus MC Comparisons

Figures 4.21 and 4.22 compare the M_{ee} distributions of smeared MC with data. Good agreement is seen in cases where there are two CC electrons. In events with at least one EC electron, the agreement is marginal; the MC energy scale is too high in CCEC events, and too low in ECEC events. A possible explanation is that the EC electron in a ECEC event will have, on average, a higher energy than the EC electron in a CCEC event with the same dielectron invariant mass. A general form of the energy response is,

$$E_{\text{meas}} = \alpha E_{\text{true}} + \beta + \gamma E_{\text{true}}^2.$$

If the EC energy response has either a negative offset (β) or positive quadratic correction (γ) relative to the MC, then the observed effect would be expected. Further study along these lines is suggested. Nevertheless, the accuracy with which the MC describes the EM energy resolution and scale is significantly better than before this work.

References

1. T. Andeen et al., Measurement of the w boson mass with 1 fb^{-1} of dø run ii data. D0 note 5893, D0, 2009
2. H. Yin, Measurement of the Forward-Backward Charge Asymmetry using $p\bar{p}$ to Z/γ^* to e^+e^- events at $\sqrt{s} = 1.96$ TeV. PhD thesis, University of Science and Technology of China, 2010
3. J.E. Gaiser, Charmonium Spectroscopy from Radiative Decays of the J/Ψ and Ψ'. PhD thesis, Stanford University, Stanford, California 94305, 1982
4. P. Gris, D0 Note 5400, D0, 2007

Chapter 5
Novel Variables for Studying the Drell-Yan Transverse Momentum

5.1 Previous Measurements

The dilepton transverse momentum, $p_T^{\ell\ell}$, distribution in $Z/\gamma^* \to \ell^+\ell^-$ production has been measured in $p\bar{p}$ collisions at $\sqrt{s} = 1.96\,\text{TeV}$ at the Fermilab Tevatron, by the CDF [1] and D0 [2–4] Collaborations. The most recent of the above measurements [3, 4] used approximately $1\,\text{fb}^{-1}$ of data. Although this represents only about one tenth of the anticipated final Tevatron data set, the precision of these measurements was already limited by experimental systematic uncertainties in the corrections for event selection efficiencies and unfolding of lepton momentum resolution. In order to unfold measured distributions for experimental resolution it is important that the chosen bin widths are not too small compared to the experimental resolution. In the low $p_T^{\ell\ell}$ region in [3, 4], the minimum bin sizes were determined by experimental resolution rather than the available data statistics. The final Tevatron data set will be an order of magnitude larger than that analysed in [3, 4]. New ideas are therefore needed in order to exploit fully the data for studying this area of physics. The ideas proposed in this chapter have been published in Refs. [5] and [6]. In Chap. 6, one of the ideas proposed in this chapter is used in an analysis with $7.3\,\text{fb}^{-1}$ and both dielectron and dimuon decay channels.

In this chapter, various alternative variables are proposed, that are sensitive to the same physics as the $p_T^{\ell\ell}$, but are less susceptible to resolution and event selection efficiency. Most of these variables are inspired by the simple fact that particle angles are better measured than particle energies/momenta.

5.2 First Idea: The a_T Variable

The a_T variable was introduced in Ref. [7], which reports on a search for anomalous production of acoplanar dilepton events with the OPAL experiment at LEP. Figure 5.1 illustrates this and other variables that are relevant to this chapter.

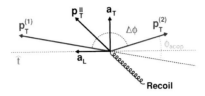

Fig. 5.1 Graphical illustration in the plane transverse to the beam direction of the variables defined in the text and used to analyse dilepton transverse momentum distributions in hadron colliders

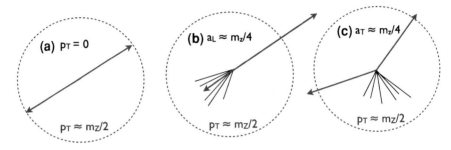

Fig. 5.2 *Left* illustration of a $Z/\gamma^* \to \ell^+\ell^-$ decay with zero $p_T^{\ell\ell}$, viewed in the transverse ($r - \phi$) plane. *Middle* a Z/γ^* decay with a moderate a_L component of the $p_T^{\ell\ell}$. *Right* a Z/γ^* decay with a moderate a_T component of the $p_T^{\ell\ell}$

Events with $\Delta\phi > \pi/2$, where $\Delta\phi$ is the azimuthal opening angle of the lepton pair, correspond to approximately 99 % of the total cross section. For such events the $p_T^{\ell\ell}$ is split into two components with respect to an event axis defined as, $\hat{t} = \left(\vec{p}_T^{\,(1)} - \vec{p}_T^{\,(2)}\right) / |\vec{p}_T^{\,(1)} - \vec{p}_T^{\,(2)}|$, where $\vec{p}_T^{\,(1)}$ and $\vec{p}_T^{\,(2)}$ are the lepton momentum vectors in the plane transverse to the beam direction. The component transverse to the event axis is denoted by a_T and the aligned component is denoted by a_L. For events with $\Delta\phi < \pi/2$ this decomposition is not useful and a_T and a_L are defined as being equal to $p_T^{\ell\ell}$.

At low $p_T^{\ell\ell}$, $\Delta\phi \sim \pi$, hence the uncertainty on a_T is approximately the uncertainty on the individual lepton p_T's multiplied by the sine of a small angle. In contrast, the uncertainty on a_L (and thus also $p_T^{\ell\ell}$) is approximately the uncertainty on the individual lepton p_T's multiplied by the cosine of a small angle. Therefore, in the low $p_T^{\ell\ell}$ region, a_T is less susceptible than $p_T^{\ell\ell}$ to the lepton p_T resolution.

Analyses of $Z/\gamma^* \to e^+e^-$ and $Z/\gamma^* \to \mu^+\mu^-$ events typically require the electrons and muons to be isolated from additional activity. This is necessary to reduce backgrounds from hadronic jets. The $p_T^{\ell\ell}$ is highly correlated with the efficiency for events to pass such isolation requirements. Figure 5.2 helps us to understand that it is actually the a_L component which is to blame for this. An event with moderate a_L also has a moderate amount of recoil activity close to one of the leptons, and is thus less likely to pass the isolation requirements. The a_T component is expected to be less correlated with isolation efficiencies, as will be demonstrated later in this

5.2 First Idea: The a_T Variable

Chapter. Trigger and geometric limitations of the detectors typically require leptons to satisfy $p_T > 15 - 25\,\text{GeV}$ and $|\eta| < 2 - 3$. The resulting acceptance depends on both the a_T and a_L components, though quite differently, as will be demonstrated later. Correlation with the kinematic acceptance is not necessarily a problem, since a measurement can (and should) be made within the acceptance.

5.3 Second Idea: Mass Ratios of a_T and $p_T^{\ell\ell}$

For $\Delta\phi \approx \pi$, a_T is given by the approximate formula

$$a_T = 2 \frac{p_T^{(1)} p_T^{(2)}}{p_T^{(1)} + p_T^{(2)}} \sin \Delta\phi,$$

and thus the fractional change in a_T with respect to a variation in, say, $p_T^{(1)}$ is given by

$$\frac{\Delta a_T}{a_T} = \frac{p_T^{(2)}}{p_T^{(1)} + p_T^{(2)}} \frac{\Delta p_T^{(1)}}{p_T^{(1)}}.$$

The dilepton invariant mass is given by

$$M_{\ell\ell} = \sqrt{2 p^{(1)} p^{(2)} (1 - \cos \Delta\theta)},$$

where $p^{(1)}$ and $p^{(2)}$ are the lepton momenta and $\Delta\theta$ is the angle between the two leptons. Thus, the fractional change in mass with respect to a variation in $p^{(1)}$ is given by

$$\frac{\Delta M_{\ell\ell}}{M_{\ell\ell}} = \frac{1}{2} \frac{\Delta p^{(1)}}{p^{(1)}}.$$

Since track angles are extremely well measured it can be taken to a very good approximation that

$$\frac{\Delta p_T^{(1)}}{p_T^{(1)}} = \frac{\Delta p^{(1)}}{p^{(1)}}.$$

The fractional change in $a_T / M_{\ell\ell}$ with respect to a variation in $p^{(1)}$ is thus given by

$$\frac{\Delta (a_T / M_{\ell\ell})}{(a_T / M_{\ell\ell})} = \frac{\Delta a_T}{a_T} - \frac{\Delta M_{\ell\ell}}{M_{\ell\ell}} = \left(\frac{p_T^{(2)}}{p_T^{(1)} + p_T^{(2)}} - \frac{1}{2} \right) \frac{\Delta p_T^{(1)}}{p_T^{(1)}}.$$

Thus the variations with $p_T^{(1)}$ in a_T and $M_{\ell\ell}$ partially cancel in the ratio, rendering $a_T/M_{\ell\ell}$ less susceptible to the effects of lepton p_T resolution than a_T. In particular, in the region of low $p_T^{\ell\ell}$ then $p_T^{(1)} \approx p_T^{(2)}$ and thus $\Delta(a_T/M_{\ell\ell}) \approx 0$. Similarly, the quantity $p_T^{\ell\ell}/M_{\ell\ell}$ is less susceptible to the effects of lepton p_T resolution than $p_T^{\ell\ell}$.

A simple example of an uncertainty in the lepton p_T scale calibration is to consider the p_T of all leptons to be multiplied by a constant factor. It can be seen trivially that in this case a_T, $p_T^{\ell\ell}$ and $M_{\ell\ell}$ are all multiplied by the same factor and that the measured $a_T/M_{\ell\ell}$ and $p_T^{\ell\ell}/M_{\ell\ell}$ are unaffected by such a scale uncertainty.

5.4 Third Idea: The ϕ_η^* Variable

A recent paper [8] has discussed the idea of using $\Delta\phi$, as an analysing variable that is sensitive to the physics of $p_T^{\ell\ell}$, and insusceptible to lepton momentum uncertainties.[1] Whilst $\Delta\phi$ is primarily sensitive to the same component of $p_T^{\ell\ell}$ as a_T, the translation from a_T to $\Delta\phi$ depends on the scattering angle θ^* of the leptons relative to the beam direction in the dilepton rest frame. Thus, $\Delta\phi$ is less directly related to $p_T^{\ell\ell}$ than a_T. As a result, $\Delta\phi$ has somewhat smaller statistical sensitivity to the underlying physics than a_T.

For $p_T^{(1)} \approx p_T^{(2)}$ it can be fairly easily shown that,

$$a_T/M_{\ell\ell} \approx \tan(\phi_{\text{acop}}/2)\sin(\theta^*),$$

where $\phi_{\text{acop}} = \pi - \Delta\phi$. This suggests that the variable,

$$\phi^* \equiv \tan(\phi_{\text{acop}}/2)\sin(\theta^*)$$

may be an appropriate alternative quantity with which to study $p_T^{\ell\ell}$.

In the analysis of hadron–hadron collisions, θ^* is commonly evaluated in the Collins–Soper frame [9]. However, θ_{CS}^* requires knowledge of the lepton momenta and is thus susceptible to the effects of lepton momentum resolution. Motivated by the desire to obtain a measure of the scattering angle that is based entirely on the measured track directions (since this will give the best experimental resolution) we propose here an alternative definition of θ^*. We apply a Lorentz boost along the beam direction such that the two leptons are back-to-back in the $r - \theta$ plane. This Lorentz boost corresponds to,

$$\beta = \tanh\left(\frac{\eta^- + \eta^+}{2}\right),$$

and yields the result

---

[1] We note that the expected distribution of $\Delta\phi$ does have a small residual sensitivity to the lepton p_T measurement. This arises from the cut on $M_{\ell\ell}$ in the event sample selection, which is affected by the lepton p_T scale and resolution.

5.4 Third Idea: The ϕ_η^* Variable

$$\cos(\theta_\eta^*) = \tanh\left(\frac{\eta^- - \eta^+}{2}\right),$$

where η^- and η^+ are the pseudorapidities of the negatively and positively charged lepton, respectively.

We consider two candidate variables,

$$\phi_{CS}^* \equiv \tan(\phi_{\text{acop}}/2)\sin(\theta_{CS}^*),$$

$$\phi^* \equiv \tan(\phi_{\text{acop}}/2)\sin(\theta_\eta^*),$$

for further evaluation in terms of their experimental resolution and physics sensitivity.

5.5 Simple Parameterised Detector Simulation

Monte Carlo events are generated using PYTHIA [10], for the process $p\bar{p} \to Z/\gamma^*$, in the e^+e^- and $\mu^+\mu^-$ decay channels, and re-weighted in dilepton $p_T^{\ell\ell}$ and rapidity, y, to match the higher order predictions of RESBOS [11]. This reweighting can be determined for a range of values of the g_2 parameter, which controls the non-perturbative effects in RESBOS (see Fig. 1.9). Electrons and muons are defined at "particle level" according to the prescription in [12], and at "detector level" by applying simple Gaussian smearing to the particle level momenta: $\delta(1/p_T) = 3 \times 10^{-3}\,\text{GeV}^{-1}$ for muons, which are measured in the tracking detectors; $\delta p/p = 0.4\sqrt{p_0/p}$ with $p_0 = 1\,\text{GeV}$, for electrons which are measured in the calorimeter. In addition, the particle angles are smeared, assuming Gaussian resolutions of 0.3×10^{-3} rad for ϕ and 1.4×10^{-3} for η. These energy, momentum, and angular resolutions roughly correspond to those in the D0 detector [13].

5.6 Scaling Factors

In the following sections, we compare the experimental resolution and physics sensitivity of the various candidate variables. In particular, we compare the variation of the resolution for each variable as a function of that variable. This comparison is facilitated by ensuring that all distributions have approximately the same scale and shape. Compared to $p_T^{\ell\ell}$ or $p_T^{\ell\ell}/M_{\ell\ell}$, all other variables are on average a factor $\sqrt{2}$ smaller (since a_T and a_L measure one component of $p_T^{\ell\ell}$). A simple multiplication by m_Z (=91.19 GeV [14]) corrects for the average $1/M_{\ell\ell}$ factor in the mass ratio and angular variables and conveniently ensures that all variables have the same units (GeV). Finally, the mean value of $\sin(\theta^*)$ is around ~0.85, and $\tan(\phi_{\text{acop}}/2)$ is scaled by this additional factor. The above factors are summarised in Table 5.1. In Sect. 5.9

Table 5.1 Scaling factors for different candidate variables

Variable	Scaling factor
$p_T^{\ell\ell}$	1
$p_T^{\ell\ell}/M_{\ell\ell}$	m_Z
a_T	$\sqrt{2}$
$a_T/M_{\ell\ell}$	$\sqrt{2}m_Z$
a_L	$\sqrt{2}$
$a_L/M_{\ell\ell}$	$\sqrt{2}m_Z$
$\tan(\phi_{\text{acop}}/2)$	$0.85\sqrt{2}m_Z$
ϕ_{CS}^*	$\sqrt{2}m_Z$
ϕ^*	$\sqrt{2}m_Z$

the distributions of the various variables (after scaling) will be shown in the context of studying efficiencies and acceptances.

5.7 Experimental Resolution for Dilepton Scattering Angle

Figure 5.3 shows the experimental resolution of $\cos(\theta_{\text{CS}}^*)$ and $\cos(\theta_\eta^*)$ in our simulation of dimuon events. The upper row of Fig. 5.3 shows events that satisfy $70 < M_{\ell\ell} < 110\,\text{GeV}$; it demonstrates that $\cos(\theta_\eta^*)$ is significantly better measured experimentally than $\cos(\theta_{\text{CS}}^*)$. This is because $\cos(\theta_\eta^*)$ is evaluated using only angular measurements, which are more precise than the momentum measurements included in the determination of $\cos(\theta_{\text{CS}}^*)$.

The variable $\cos(\theta_\eta^*)$ is used in the definition of $\phi^* = \tan(\phi_{\text{acop}}/2)\sin(\theta_\eta^*)$ in Sect. 5.4 above. As an aside, we note in addition that a precise determination of the dilepton centre of mass scattering angle that is free from the effects of lepton momentum resolution can also find application in the determination of the forward–backward charge asymmetry of dilepton production at hadron colliders. The experimental resolution in $\cos(\theta_{\text{CS}}^*)$ becomes particularly significant in the dimuon channel for very high values of $M_{\ell\ell}$ for which the lepton momentum resolution is poorest. This is illustrated in the lower row of Fig. 5.3, which shows the experimental resolution of $\cos(\theta_{\text{CS}}^*)$ and $\cos(\theta_\eta^*)$ in events that satisfy $500 < M_{\ell\ell} < 600\,\text{GeV}$. The advantage of using $\cos(\theta_\eta^*)$ for high mass events is even larger than was the case for $70 < M_{\ell\ell} < 110\,\text{GeV}$.

5.8 Experimental Resolution for Variables Related to $p_T^{\ell\ell}$

Figure 5.4 compares separately for our dimuon and dielectron simulations, the RMS resolution of the candidate variables as a function that variable (at particle level). All variables are scaled by the factors in Table 5.1.

5.8 Experimental Resolution for Variables Related to $p_T^{\ell\ell}$

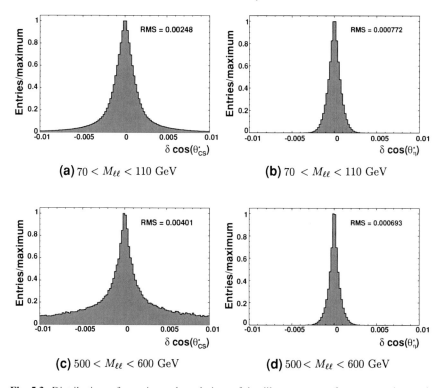

Fig. 5.3 Distributions of experimental resolutions of the dilepton centre of mass scattering angle for events satisfying $70 < M_{\ell\ell} < 110$ GeV (*upper row*) and $500 < M_{\ell\ell} < 600$ GeV (*lower row*). Part labels (**a**) and (**c**) show $\cos(\theta_{CS}^*)$. Part labels (**b**) and (**d**) show $\cos(\theta_\eta^*)$

The following observations are made:

- $a_T/M_{\ell\ell}$ is significantly better measured than a_T, over the entire range.
- Similarly, $p_T^{\ell\ell}/M_{\ell\ell}$ is significantly better measured than $p_T^{\ell\ell}$.
- Over the full range, a_T and $a_T/M_{\ell\ell}$ perform better than $p_T^{\ell\ell}$ and $p_T^{\ell\ell}/M_{\ell\ell}$ respectively.
- Compared to $a_T/M_{\ell\ell}$, ϕ_{CS}^* has significantly better resolution, and ϕ^* better still.
- The most precisely measured variable is $\tan(\phi_{\mathrm{acop}}/2)$, since it is determined only from the azimuthal angles of the leptons, whereas the uncertainty on ϕ^* includes also the uncertainties on the measured lepton pseudorapidities.

Since the discussion in Sect. 5.3 is only approximate, we have investigated empirically various other possible scalings of a_T with $M_{\ell\ell}$ (with the appropriate scale factor applied, as above). These are illustrated in Fig. 5.5. As expected, it can be seen that when compared to the other variables considered in Fig. 5.5, $a_T/M_{\ell\ell}$ has the best experimental resolution for all $p_T^{\ell\ell}$ and irrespective of whether (a) tracker-like or (b) calorimeter-like resolution in the lepton momenta is simulated.

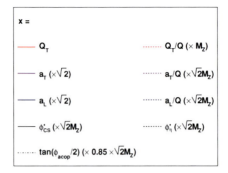

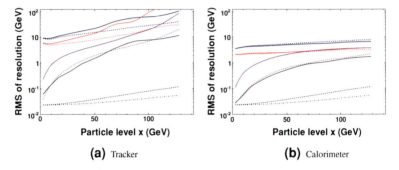

Fig. 5.4 The RMS resolution of each candidate variable, as a function of that variable (scaled by the factors described in the text). Results are presented both for **a** tracker-like and **b** calorimeter-like resolution in the lepton momenta

5.9 Acceptance and Efficiency

As discussed in Sect. 5.2 the a_T component of the $p_T^{\ell\ell}$ is expected to be less correlated with isolation efficiencies than the a_L component. In addition, the "sculpting" effects of kinematic acceptance on the various variables will be very different. The kinematic acceptances are compared for the different variables and the following cuts:

1. Both leptons satisfy $p_T > 20\,\text{GeV}$,
2. Both leptons satisfy $|\eta| < 2$,
3. Both leptons satisfy $p_T > 20\,\text{GeV}$ and $|\eta| < 2$.

These requirements are made at the generator level. Figure 5.6 shows for $p_T^{\ell\ell}$, a_T, a_L, $a_T/M_{\ell\ell}$, and ϕ_η^*, the p_T cut acceptance. The acceptance decreases rapidly as a function of a_L, since a larger value of a_L tends to push one lepton to lower p_T. For a_L above roughly $m_Z/2$, the dependence plateaus, since a further increase in a_L now increases the p_T of both leptons. It is not surprising that for a_T, the dependence is

5.9 Acceptance and Efficiency

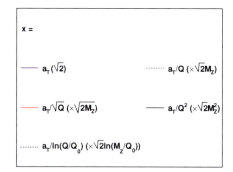

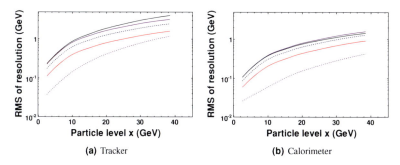

Fig. 5.5 The RMS resolution of the variables a_T, $a_T/M_{\ell\ell}$, $a_T/M_{\ell\ell}^{1/2}$, $a_T/\ln(M_{\ell\ell}/Q_0)$, and $a_T/M_{\ell\ell}^2$, as a function of that variable (scaled by the factors described in the text). Results are presented both for **a** tracker-like and **b** calorimeter-like resolution in the lepton momenta. The Q_0 parameter is simply introduced to remove the mass dimension in the logarithm

in the opposite direction. Once a_T is above $m_Z/2$ (multiplied by the scaling factor in Fig. 5.6), the p_T cut acceptance is 100 %. For $p_T^{\ell\ell}$, the dependence is somewhere in between that for a_T and a_L.

Figure 5.7 shows for $p_T^{\ell\ell}$, a_T, a_L, $a_T/M_{\ell\ell}$, and ϕ_η^*, the η cut acceptance. The dependence is rather similar for a_T, a_L, $p_T^{\ell\ell}$, and $a_T/M_{\ell\ell}$. This is easily understood, since an increase in the $p_T^{\ell\ell}$ will tend to push the leptons more into the central region of the detector. Figure 5.8 shows the acceptance for the combination of p_T and η cuts. This acceptance is relatively flat for ϕ_η^* compared to the other variables.

In order to study the isolation efficiency dependence, we define the lepton isolation variable $\mathcal{I}_{\text{gen}}^{\text{hc4}}$ as the p_T sum of all particles apart from neutrinos within a hollow cone of $0.1 < \Delta R < 0.4$ around the lepton. Our isolation requirement is that both leptons must have $\mathcal{I}_{\text{gen}}^{\text{hc4}}/p_T < 0.1$. This is comparable to the requirements that we made in selecting a sample of $Z/\gamma^* \to e^+e^-$ events in Chap. 4. Figures 5.9 and 5.10 show the isolation efficiencies without and with kinematic cuts (p_T and η cuts described above) respectively. The isolation efficiency shows a strong dependence on a_L, which is easily understood (see earlier discussion in Sect. 5.2). The dependence on a_T is significantly smaller, and is reduced after having imposed the lepton p_T and η

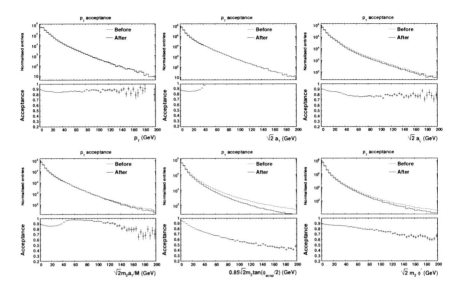

Fig. 5.6 Acceptance (for the requirement of both leptons with $p_T > 15\,\text{GeV}$) for the various candidate variables. The *upper half* of the plots show the distributions before and after acceptance, and the *lower halves* show the acceptances

acceptance. The dependence on ϕ^*_η is similar to that on a_T which is expected since they are primarily sensitive to the same component of the $p_T^{\ell\ell}$.

5.10 Sensitivity to the Physics

Figure 5.11 shows the particle level, normalised distributions of a_T, $m_Z a_T / M_{\ell\ell}$ and $\ln(m_Z/Q_0)a_T / \ln(M_{\ell\ell}/Q_0)$ (with $Q_0 = 1\,\text{GeV}$) for three ranges of $M_{\ell\ell}$. The Q_0 parameter is simply introduced to remove the mass dimension in the logarithms. We see that a_T has a mild dependence on $M_{\ell\ell}$, while dividing by $M_{\ell\ell}$ over corrects this dependence. In this respect, we observe that the distribution in $a_T / \ln(M_{\ell\ell}/Q_0)$ has a smallest dependence on $M_{\ell\ell}$. This is to be expected since QCD predicts the mean $p_T^{\ell\ell}$ grows logarithmically with $M_{\ell\ell}$ [15].

Experimental measurements of Z/γ^* production are typically made over a fairly wide bin in $M_{\ell\ell}$ around m_Z (e.g., 70–110 GeV). One potential concern with measurements of $a_T/M_{\ell\ell}$ and $p_T^{\ell\ell}/M_{\ell\ell}$ is that the increased correlation with $M_{\ell\ell}$ demonstrated in Figure 5.11 might degrade the sensitivity to the underlying physics. Since ϕ^* behaves approximately as $\phi^* \approx a_T/M_{\ell\ell}$, a similar degradation in the physics sensitivity of ϕ^* may similarly be expected. In this respect, $a_T / \ln(M_{\ell\ell}/Q_0)$ is a more suitable variable than $a_T/M_{\ell\ell}$ for studying the boson $p_T^{\ell\ell}$ distribution over a wide range in $M_{\ell\ell}$. However it has poorer experimental resolution, as was demonstrated in Fig. 5.5.

5.10 Sensitivity to the Physics

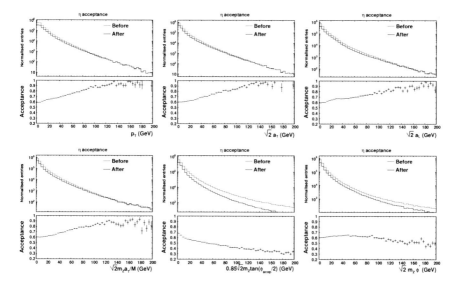

Fig. 5.7 Acceptance (for the requirement of both leptons with $|\eta| < 2$) for the various candidate variables. The *upper half* of the plots show the distributions before and after acceptance, and the *lower halves* show the acceptances

In order to compare the physics sensitivity of the different candidate variables, we run pseudo-experiments to fit for the value of the parameter g_2, which determines the width of the low $p_T^{\ell\ell}$ region in RESBOS (see Chap. 1). Events must meet the following requirements: both leptons with $p_T > 15\,\text{GeV}$ and $|\eta| < 2$, as calculated at detector level. Of these events, 1M are assigned as pseudo-data and the remaining events are used to build g_2 templates. All variables are scaled by the factors listed in Table 5.1, such that the same binning (30 equal width bins in the range 0–30 GeV) can be used. A minimum χ^2 fit determines the statistical sensitivity of each variable to the value of g_2.

The g_2 parameter mostly governs the shape of the distribution for $p_T < 25\,\text{GeV}$. We similarly fit for a parameter K_{p_T} which weights events with (particle level) $p_T^{\ell\ell} > 25\,\text{GeV}$, by $K_{p_T}(p_T^{\ell\ell} - 25)$, and approximately represents the differences between predictions at NLO and NNLO discussed in [4]. Again, after applying the appropriate scaling factors from Table 5.1, the same binning can be used for each variable.[2]

The results of the fits to g_2 and K_{p_T} are presented in Tables 5.2 and 5.3, respectively. Results are given separately for particle-level (dimuon) and detector-level (tracker and calorimeter). For both $a_T/M_{\ell\ell}$ and $p_T^{\ell\ell}/M_{\ell\ell}$, the statistical sensitivities

---

[2] The first bin is of width 5 GeV (with lower edge at zero) and each consecutive bin is 2 GeV wider than the last. Ten such bins give an upper edge of the last bin at 140 GeV and the fit includes the overflow bin from 140 GeV to ∞.

76   5 Novel Variables for Studying the Drell-Yan Transverse Momentum

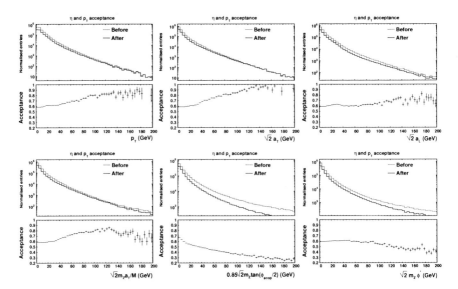

Fig. 5.8 Acceptance (for the requirement of both leptons with $p_T > 15$ GeV and $|\eta| < 2$) for the various candidate variables. The *upper half* of the plots show the distributions before and after acceptance, and the *lower halves* show the acceptances

are essentially the same as those for a_T and $p_T^{\ell\ell}$ respectively. Thus the effect of the additional $M_{\ell\ell}$ dependence is shown to be negligible.

The approximately 5% poorer sensitivity of $\tan(\phi_{\text{acop}}/2)$, compared to $a_T/M_{\ell\ell}$, demonstrates the $\sin(\theta^*)$ ambiguity of the former. The additional factor $\sin(\theta^*)$, present in ϕ^* and ϕ_{CS}^* actually recovers the sensitivity to the same level as a_T. In addition, the ϕ^* variable, which was shown in Sect. 5.8 to have the best experimental resolution (except for $\tan(\phi_{\text{acop}}/2)$), performs similarly to ϕ_{CS}^* in terms of physics sensitivity.

Of course, the results presented in Tables 5.2 and 5.3, represent only the statistical sensitivity of the considered variables when compared to $p_T^{\ell\ell}$. A variable that is less sensitive to resolution and efficiency effects will inevitably make such a measurement less sensitive to the precise modelling of these effects in the MC, thus reducing the systematic uncertainties.

5.11 Discussion on the Different Variables

The first idea is to use the a_T variable introduced in Ref. [7]. The a_T component of the $p_T^{\ell\ell}$ has the advantage of being less correlated with isolation efficiencies, and, in the region of low $p_T^{\ell\ell}$ having a better experimental resolution than the $p_T^{\ell\ell}$. This idea has little or no benefit for studying the high $p_T^{\ell\ell}$ tail.

5.11 Discussion on the Different Variables

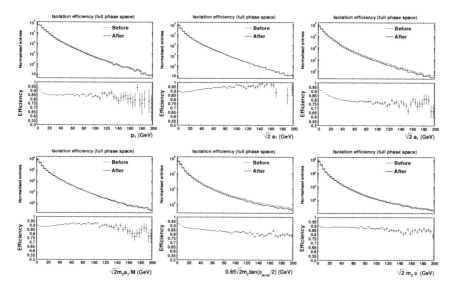

Fig. 5.9 Isolation efficiency (isolation requirements described in the *text*) for the various candidate variables. The *upper half* of the plots show the distributions before and after isolation cuts, and the *lower halves* show the efficiencies. No kinematic cuts have been applied

Table 5.2 Statistical sensitivity (in %) on the parameter g_2 from fits to the distributions of different of variables

Variable	Particle level	Calorimeter	Tracker
$p_T^{\ell\ell}$	0.65	0.94	1.41
$p_T^{\ell\ell}/M_{\ell\ell}$	0.65	0.94	1.40
a_T	1.00	1.00	1.00
$a_T/M_{\ell\ell}$	1.00	1.01	1.00
a_L	1.21	2.35	4.74
$\tan(\phi_{acop}/2)$	1.04	1.05	1.04
ϕ_{CS}^*	1.00	1.00	0.99
ϕ^*	1.00	1.00	0.99

For details see text

An improvement of this idea is to take the ratio to the measured dilepton invariant mass, i.e. $a_T/M_{\ell\ell}$. Mis-measurements of either of the lepton p_Ts are partially cancelled in this ratio, and uncertainties in absolute momentum scale are almost totally cancelled. The $a_T/M_{\ell\ell}$ variable has a similar correlation with isolation efficiencies to that of a_T, but has significantly better resolution in the region of high $p_T^{\ell\ell}$. The physics sensitivity is not degraded by the division by the mass. The same idea actually works to some extent with $p_T^{\ell\ell}/M_{\ell\ell}$ compared to the $p_T^{\ell\ell}$.

In order to fully avoid the problem of lepton p_T mis-measurement, we need to build a variable exclusively from angles. An obvious candidate is the azimuthal opening angle between the two leptons, $\Delta\phi$. Whilst the $\Delta\phi$ is primarily sensitive to

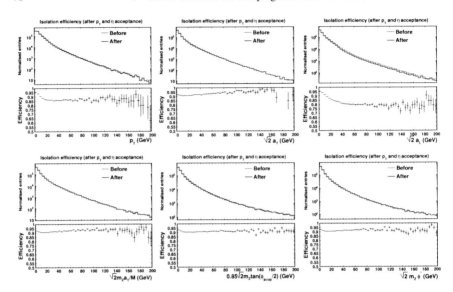

Fig. 5.10 Isolation efficiency (isolation requirements described in the text) for the various candidate variables. The upper half of the plots show the distributions before and after isolation cuts, and the lower halves show the efficiencies. The requirements, $p_T > 15\,\text{GeV}$ and $|\eta| < 2$ have been imposed on both leptons

Table 5.3 Statistical uncertainty (in %) on the parameter K_{p_T} (as defined in the text) from fits to the distributions of different of variables

Variable	Particle level	Calorimeter	Tracker
$p_T^{\ell\ell}$	1.65	1.67	1.82
$p_T^{\ell\ell}/M_{\ell\ell}$	1.66	1.67	1.78
a_T	1.92	1.92	1.96
$a_T/M_{\ell\ell}$	1.92	1.92	1.94
a_L	1.98	2.02	2.34
$\tan(\phi_{\text{acop}}/2)$	1.96	1.96	1.98
ϕ_{CS}^*	1.88	1.88	1.90
ϕ^*	1.87	1.87	1.92

For details see text

the a_T component of the $p_T^{\ell\ell}$, the translation from $\Delta\phi$ to a_T depends on the scattering angle. The ϕ_η^* variable corrects for this, based on an estimator of the scattering angle that uses only angular information. The physics sensitivity of ϕ_η^* is comparable to that of a_T, and has similar advantages in terms of efficiency correlations, but the variable has essentially perfect experimental resolution over the entire range of $p_T^{\ell\ell}$. It is therefore concluded that ϕ_η^* is the optimum variable for studying the $p_T^{\ell\ell}$ physics in a detector with limited resolution in lepton p_T.

5.11 Discussion on the Different Variables 79

Fig. 5.11 Comparison of particle level distributions of a_T, $m_Z a_T/M_{\ell\ell}$ and $\ln(m_Z/Q_0)a_T/\ln(M_{\ell\ell}/Q_0)$ for three ranges of $M_{\ell\ell}$

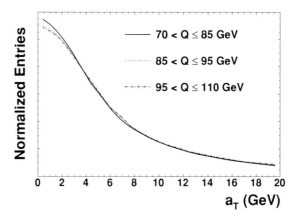
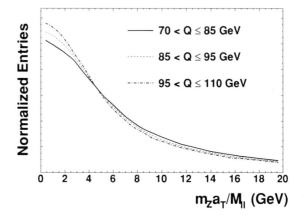
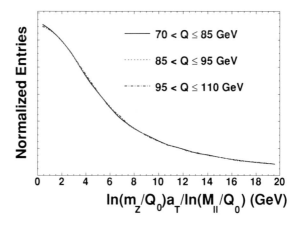

References

1. T. Affolder et al., Phys. Rev. Lett. **84**, 845 (2000)
2. B. Abbott et al., Phys. Rev. D **61**, 032004 (2000)
3. V.M. Abazov et al., Phys. Rev. Lett. **100**, 102002 (2008)
4. V.M. Abazov et al., Phys. Lett. B **693**, 522 (2010)
5. M. Vesterinen, T.R. Wyatt. A novel technique for studying the Z boson transverse momentum distribution at hadron colliders. Nucl. Instr. Meth. Phys. Res. A **602**, 432–437 (2009)
6. A. Banfi, S. Redford, M. Vesterinen, P. Waller, T.R. Wyatt, Optimisation of variables for studying dilepton transverse momentum distributions at hadron colliders. Eur. Phys. J. C **71**, 1600 (2011)
7. K. Ackerstaff, Eur. Phys. J. C **4**, 47 (1998)
8. M. Boonekamp, M. Schott. arXiv:1002.1850v1 [hep-ex] (2010)
9. J. Collins, D. Soper, Phys. Rev. D **16**, 2219–2225 (1997)
10. T. Sjostrand, Comput. Phys. Comm. **135**, 238 (2001)
11. C. Balazs, C.-P. Yuan, Phys. Rev. D **56**, 5558–5583 (1997) (We use the CP version of the code and grid files)
12. J.M. Butterworth et al., The Tools and Monte Carlo Working Group Summary Report from the Les Houches 2009 Workshop on TeV Colliders. In: Les Houches 2009 Tools and Monte Carlo working group (2010). arXiv:1003.1643v1 [hep-ph]
13. V.M. Abazov et al., The upgraded D0 detector. Nucl. Instr. Meth. Phys. Res. A **565**(2), 463–537 (2006)
14. K. Nakamura et al., J. Phys. G **37**, 075021 (2010)
15. J. Collins, D. Soper, G. Sterman. Transverse momentum distribution in Drell-Yan pair and W and Z boson production. Nucl. Phys. B **250**, 199–224 (1985)

Chapter 6
Measurement of the Drell-Yan ϕ_η^* Distribution

We now have a variable, ϕ_η^*, that allows us to fully exploit the huge $Z/\gamma^* \to e^+e^-$ and $Z/\gamma^* \to \mu^+\mu^-$ event samples in studying the $p_T^{\ell\ell}$ physics. Despite the relatively poor muon momentum resolution, similar precision can be achieved in the dimuon and dielectron channels. Compared to previous analyses that only considered the dielectron channel, the ability to compare consistency between the two channels can only help to increase confidence in the accuracy of the measurement. The large event samples also allow us to study in detail the rapidity dependence of the $p_T^{\ell\ell}$. The measurement strategy is as follows:

- Measure the shape of the ϕ_η^* distribution, $(1/\sigma)(d\sigma/d\phi_\eta^*)$.
- Use both dielectron and dimuon decay channels.
- Measure in three bins of rapidity: $|y| < 1$, $1 < |y| < 2$, and (dielectron only) $|y| > 2$.
- The distributions are unfolded to the measurable phase space, and are not corrected for final state radiation.
- The measured distributions are compared to predictions from RESBOS [1][1] with QED radiative corrections from PHOTOS [2].

6.1 The Observables

Following the prescription in [3] to minimise the model dependence of the measurement, the data are unfolded to the measurable phase space and particle level electron and muons definitions that mimic the way in which the particles are reconstructed experimentally. Due to the different detector acceptances, measurement techniques (calorimeter vs. tracker), and effects final state radiation, there is no model independent way of combining the electron and muon channels. Therefore we choose *not to combine the two channels*. Rather, the dielectron and dimuon data are separately

---

[1] We use the CP version of the code and grid files.

M. Vesterinen, *Z Boson Transverse Momentum Distribution, and ZZ and WZ Production*, Springer Theses, DOI: 10.1007/978-3-642-30788-1_6, © Springer-Verlag Berlin Heidelberg 2012

compared to predictions from RESBOS with the appropriate FSR corrections from PHOTOS and kinematic cuts.

In addition to a dilepton invariant mass between 70 and 110 GeV, the particle level kinematic cuts are as follows: $|\eta| < 3$ and $p_T > 20$ GeV for the dielectron channel, and $|\eta| < 2$ and $p_T > 15$ GeV for the dimuon channel. Additionally, the "ICR" region $1.1 < |\eta| < 1.5$ is excluded at the particle level.

Particle level electrons are defined as the four-vector sum of all EM particles within a cone of $\Delta \mathcal{R} < 0.2$ around a seed electron. Particle level muons are defined as the muon particle *after* QED final state radiation. These definitions roughly follow the way in which the kinematic properties of electrons and muons are reconstructed in the calorimeter and tracker respectively.

6.2 Event Selection

6.2.1 Event Selection Strategy

Our objective is to select a relatively background free sample of Z/γ^* events, with maximum possible acceptance × efficiency, whilst adhering to the following:

- Require all leptons to be matched to central tracks, which have good angular resolution.
- Restrict to single-lepton and dilepton triggers, since lepton+jet triggers are expected to bias events towards larger $p_T^{\ell\ell}$. Correcting for this bias would introduce a source of systematic uncertainty.
- Limit the inefficiencies in the CC ϕ-gaps and the muon system inter-octant gaps. The ϕ_η^* distribution is particularly affected by regions of inefficiency that are back-to-back in ϕ. Figure 6.1 shows, separately for electrons and muons, how the acceptance modulates as a function of ϕ_η^* when neither lepton is allowed to be within one of these gaps.

6.2.2 Data Sample and Skims

This analysis includes the following data taking runs: 151817–215670 (Run IIa), 221698–261343 (Run IIb1–Run IIb3). The last run was recorded on the 20th May 2010. Most D0 analyses reject certain periods of data taking that are flagged as bad by the D0 data quality group. In order to maximise our available event yields, we make almost no such requirements, apart from removing events that are flagged as having severe calorimeter noise. Since this is only a *shape* measurement, it is rather insensitive to operational problems that affect the absolute efficiency, e.g., part of

6.2 Event Selection

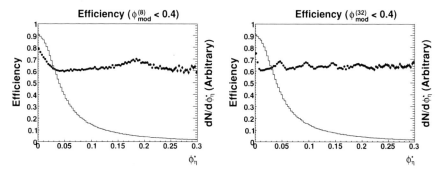

Fig. 6.1 The efficiency for $Z/\gamma^* \to \ell^+\ell^-$ events to have both leptons outside the (*left*) muon octant-gaps and the (*right*) CC ϕ-gaps, as a function of ϕ_η^*. The ϕ_η^* distributions (*red*) are overlaid for illustration. Although we *do not* make such fiducial cuts in this analysis, there are unavoidable inefficiencies in the central calorimeter and inter-octant ϕ-gaps, which will lead to similar (though smaller) efficiency dependencies

the inner tracker being switched off. After the minimal data quality requirements, the available integrated luminosity is approximately $7.3\,\text{fb}^{-1}$.

6.2.3 Monte Carlo Samples

Signal ($Z/\gamma^* \to e^+e^-$ and $Z/\gamma^* \to \mu^+\mu^-$), and EW background ($Z/\gamma^* \to \tau^+\tau^-$, $WW \to \ell^+\nu\ell^-\bar{\nu}$, and $W \to l\nu$+jet) events are generated using PYTHIA [4]. Additional samples of $t\bar{t}$ events are generated using ALPGEN [5] with showering and hadronisation by PYTHIA.

6.2.4 Common Dielectron and Dimuon Requirements

- Dilepton invariant mass between 70 and 110 GeV.
- Primary vertex within 80 cm of the detector centre along the beam direction.
- Lepton central tracks must have opposite electric charge.

6.2.5 Dielectron Event Selection

Dielectron candidate events are required to have fired any of the single-EM triggers, and contain two reconstructed electrons that are matched to central tracks, at least one of which is matched to a single-EM trigger. We require $|\Delta z| < 3\,\text{cm}$, where Δz

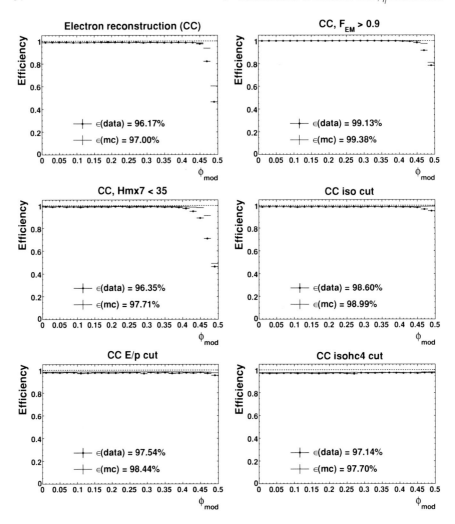

Fig. 6.2 *Top left* CC electron cluster reconstruction efficiency as a function of ϕ_{mod}. The other subfigures show the efficiencies for CC clusters to pass various other quality criteria

is the difference in the z coordinates of the positions of closest approach of the two tracks in the plane transverse to the beam centroid, and the electron candidates must meet the following additional requirements:

- Within the CC, or a restricted part of the EC ($|\eta_{\text{det}}| \leq 3.0$),
- $E_T > 20\,\text{GeV}$,
- $\mathcal{I}_{\text{cal}}^{\text{EM}} < 0.10$,
- $\mathcal{I}_{\text{trk}}^{\text{hc4}} < 2.5\,\text{GeV}$.

6.2 Event Selection

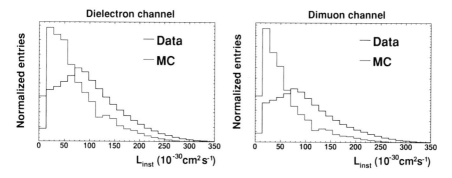

Fig. 6.3 Comparison of the normalised instantaneous luminosity distributions in data and MC, after selection cuts

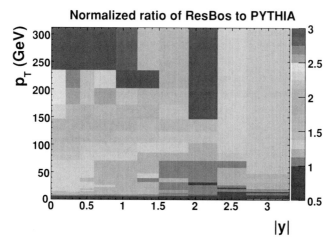

Fig. 6.4 Normalised ratio of RESBOS [1] to PYTHIA [4], in the p_T versus $|y|$ distribution of $Z/\gamma^* \to \ell^+\ell^-$ events

- CC electrons are required to satisfy $0.25 < E_{cal}/p_{trk} < 4$.
- EC electrons are required to satisfy $\chi^{2(8)}_{EM} < 40$.

These selection cuts, in particular for the CC, are chosen to maximise the efficiency in the ϕ-gaps. The efficiencies for CC electrons to pass cuts on $\mathcal{F}_{EM}$ and $\chi^{2(7)}_{EM}$ are strongly ϕ_{mod} dependent as shown in Fig. 6.2. Cuts on these variables are therefore avoided, and instead we rely more on the E_{cal}/p_{trk}, track isolation, and calorimeter isolation variables, which introduce significantly less ϕ_{mod} dependence as shown in Fig. 6.2. Unfortunately, the efficiency to simply reconstruct an EM cluster is reduced by up to 50 % in the ϕ-gaps as shown in the top left of Fig. 6.2. This inefficiency is also rather poorly modelled by the simulation—differing by up to 10 %. For the EC, the E_{cal}/p_{trk} cut is simply replaced by an $\chi^{2(8)}_{EM}$ cut. Although there is no ϕ-gap

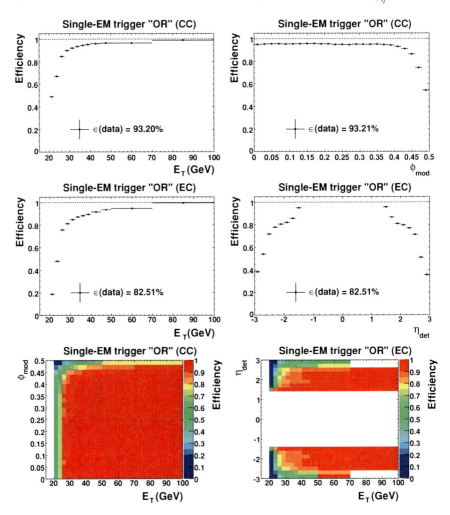

Fig. 6.5 Efficiency for the OR of all single-EM triggers

problem in the EC, we found that backgrounds can be kept under control without cutting on the EM fraction. A detailed study of efficiencies is presented in Sect. 6.3.

6.2.6 Dimuon Event Selection

Candidate dimuon events are required to have fired any one of the single-muon triggers, and meet the following offline requirements:

6.2 Event Selection

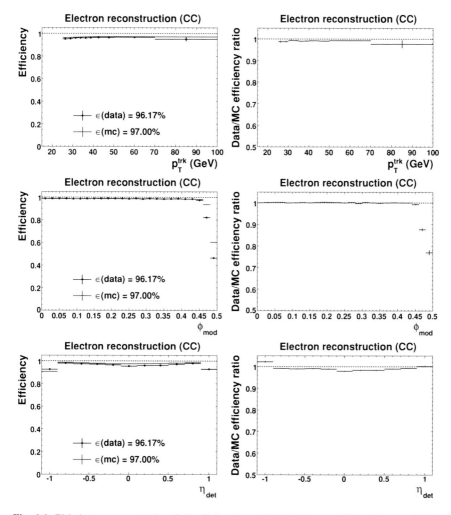

Fig. 6.6 EM cluster reconstruction (*left*) efficiencies and (*right*) data-vs-MC scale factors for the CC region

- **Muon requirements** Events must contain at least one "trigger" muon with track segments of wire and scintillator hits in the A and B/C layers (nseg = 3), and matched to a central track with $p_T > 15\,\text{GeV}$, The trigger muon must be matched to all three levels of a single-muon trigger. Events must contain a second muon with nseg $\geq$ 0, and again matched to a ($p_T > 15\,\text{GeV}$) central track. All tracks must meet the requirement, $\chi^2/\text{dof} < 9.5$. In addition, the two tracks must share a common origin, with $\Delta z < 2 - n_{\text{SMT}}/2$, where Δz is measured in cm, and n_{SMT} is the number of tracks that have at least one hit in the SMT.

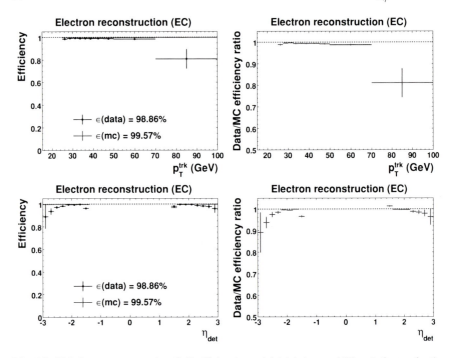

Fig. 6.7 EM cluster reconstruction (*left*) efficiencies and (*right*) data-vs-MC scale factors for the EC region

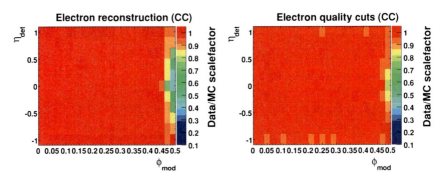

Fig. 6.8 Data-vs-MC efficiency scale factors versus ϕ_{mod} and η_{det}, for (*left*) cluster reconstruction, and (*right*) higher level quality cuts (with respect to track matched *reco* electrons)

- **Isolation requirements** Nseg = 3 muons are required to satisfy $(\mathcal{I}_{\text{cal}} - 0.005\mathcal{L}_{\text{inst}})/p_T < 0.3$ and $\mathcal{I}_{\text{trk}}/p_T < 0.15$. Nseg < 3 muons are required to satisfy $(\mathcal{I}_{\text{cal}} - 0.005\mathcal{L}_{\text{inst}})/p_T < 0.2$ and $\mathcal{I}_{\text{trk}}/p_T < 0.10$. The calorimeter isolation thus acquires a simple correction for the effect of pile-up, by subtracting a $\mathcal{L}_{\text{inst}}$ dependent term.

6.2 Event Selection

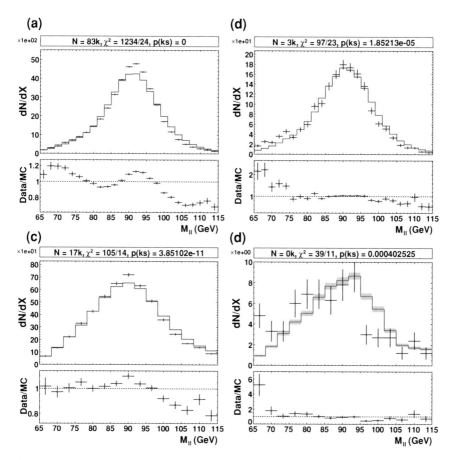

Fig. 6.9 Data-vs-MC comparison of the tag-and-probe invariant mass distributions separately for events where the probe passes/fails the studied requirement. The *black points* with error bars are the data, and the *red histogram* is the simulation. **a** CC cluster (pass). **b** CC cluster (fail). **c** EC cluster (pass). **d** EC cluster (fail)

- **Geometric cosmic rejection** The requirement, $|\eta_1 + \eta_2| > 0.02$, rejects cosmic ray muons. Although the pseudo-acolinearity, defined as $\alpha_{aco} = |\eta_1 + \eta_2| + ||\phi_1 - \phi_2| - \pi|$, is more discriminating against cosmic ray muons, we nevertheless prefer to cut on $|\eta_1 + \eta_2|$. Zero α_{aco} corresponds to zero ϕ_η^*. Hence a cut on α_{aco}, could potentially bias the measurement. A requirement on the distance of closest approach of the tracks in the plane transverse to the beam centroid, r_{dca}, further reduces cosmic contamination. We require that the mean r_{dca} of the two muons is less than 0.006, 0.015, 0.1 cm, for events with $n_{SMT} = 2, 1, 0$, respectively.

90   6   Measurement of the Drell-Yan ϕ_η^* Distribution

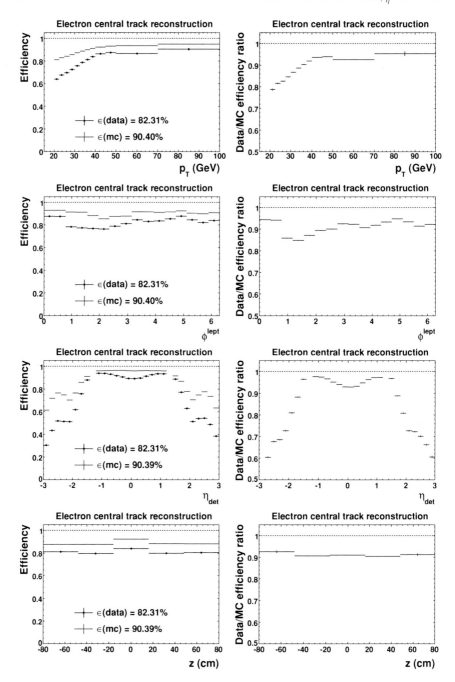

Fig. 6.10 Electron tracking (*left*) efficiency and (*right*) data-vs-MC scale factors

6.3 Corrections to the Fully Simulated Monte Carlo Events 91

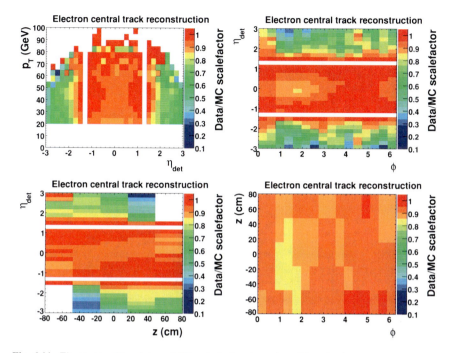

Fig. 6.11 Electron tracking data-vs-MC scale factors in various 2D projections

6.3 Corrections to the Fully Simulated Monte Carlo Events

As mentioned in Chap. 3, the standard MC simulation suffers from various inaccuracies. In this section, various corrections to the simulated MC events are detailed. Many of these corrections are determined specifically for this analysis.

6.3.1 Instantaneous Luminosity Profile

Figure 6.3 shows that, in both channels, the instantaneous luminosity profiles differ significantly between data and MC. In order to maximise *effective* MC statistics, we choose *not* to re-weight the profile in MC to that of data and instead absorb any effects on efficiency into data-vs-MC scale factors determined in this section.

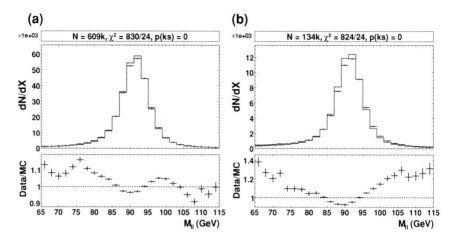

Fig. 6.12 Data-vs-MC comparison of the tag-and-probe invariant mass distributions separately for events where the probe passes/fails the studied requirement. The *black points* with error bars are the data, and the *red histogram* is the simulation. **a** Electron tracking (pass). **b** Electron tracking (fail)

6.3.2 Generator Level Physics Re-Weightings

Events are re-weighted at the generator level (propagator level Z/γ^* particle) to the predictions of RESBOS [1], with $g_2 = 0.68\,\text{GeV}^2$, in two dimensions (p_T, and y). The grid files used to generate the RESBOS events include a NLO $\to$ NNLO K-factor. This re-weighting can be performed for a range of g_2 values, which will form the basis of various closure tests. Figure 6.4 shows the ratio of normalised RESBOS and PYTHIA distributions in p_T and $|y|$, which is used in this re-weighting. Since we have insufficient statistics at high p_T ($p_T > 250$ GeV) and high $|y|$ ($|y| > 2$), we do the following to the ratio histogram: for each bin in $|y|$, we loop over the bins in (increasing) p_T, up to and including the overflows, and once we reach a bin with insufficient entries, we set the value of that bin, to the same value as the last reasonable bin. In addition, when applying the reweighting (though not in Fig. 6.4), we define a minimum weight of 0.2, and a maximum weight of 3.

6.3.3 Electron Energy and Muon Momentum Smearing

Electron energies receive the treatment derived in Chap. 4. Muon track momenta are smeared to better describe the resolution observed in data [6].

6.3.4 Track ϕ and η Smearing

The track ϕ and η resolutions are found to be better in the simulation than in the data. Appropriate Gaussian oversmearing parameters are derived for tracks with and

6.3 Corrections to the Fully Simulated Monte Carlo Events

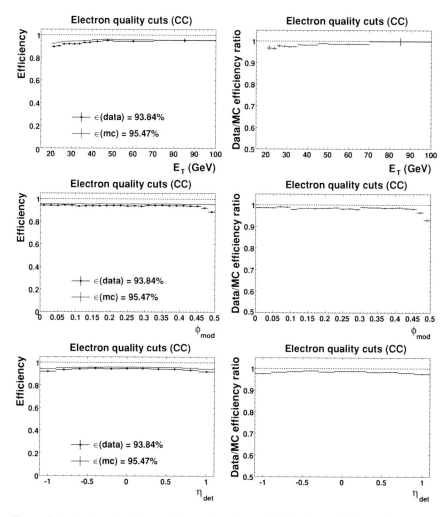

Fig. 6.13 Higher level quality cut (*left*) efficiencies and (*right*) data-vs-MC scale factors, with respect to track matched *reco* electrons in the CC

without SMT hits, in ϕ and η. In data, the resolutions are determined using cosmic ray muon events, and these are compared to the resolutions in MC. More details can be found in Ref. [7].

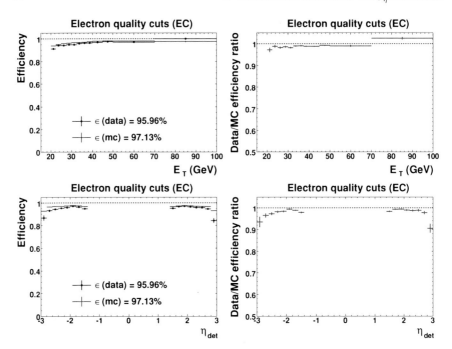

Fig. 6.14 Higher level quality cut (*left*) efficiencies and (*right*) data-vs-MC scale factors, with respect to track matched *reco* electrons in the EC

6.3.5 Electron Track p_T Smearing

It is important that the electron track p_T resolution and scale is well described by the simulation. The event selection includes an E_{cal}/p_{trk} cut for CC electrons; improved modelling of this variable will reduce the size of the data-vs-MC efficiency scale factors. Furthermore, the measurement of electron reconstruction efficiency requires a selection of $Z/\gamma^* \to e^+e^-$ events in which the kinematics of one electron must be determined from the central track. A Gaussian smearing of width $0.0025\,\text{GeV}^{-1}$ in $1/p_T$ is applied to electron tracks. In addition, the momenta are scaled by a factor of 0.998. These parameters are adjusted by eye, such that the simulation better describes the invariant mass distribution of $Z/\gamma^* \to e^+e^-$ candidates in which the central track p_T is used for one of the electrons.

6.3.6 Local Muon p_T Smearing

It is important that the local muon p_T resolution and scale is well described, for the purposes of measuring the central tracking efficiency of muons. A Gaussian

6.3 Corrections to the Fully Simulated Monte Carlo Events 95

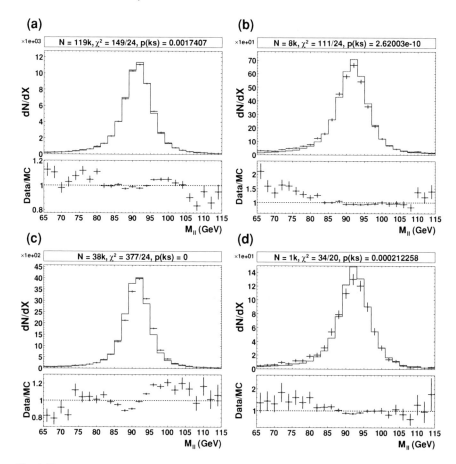

Fig. 6.15 Data-vs-MC comparison of the tag-and-probe invariant mass distributions separately for events where the probe passes/fails the studied requirement. The *black points* with error bars are the data, and the *red histogram* is the simulation. **a** CC EMID (pass). **b** CC EMID (fail). **c** EC EMID (pass). **d** EC EMID (fail)

smearing in $1/p_T$ of width $0.005\,\text{GeV}^{-1}$, and scaling factor of 0.98 are applied to all local muon momenta. These parameters are simply adjusted by eye, such that the simulation better describes the invariant mass distribution calculated using the local muon momentum from one of the muons. Example invariant mass distributions are shown in the following sections which describe the measurement of various efficiencies, including central track matching. The local muon p_T measurement is needed to select dimuon events where only one muon is required to have a central track match.

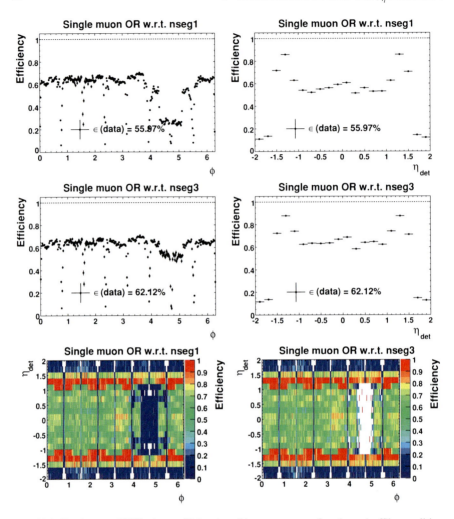

Fig. 6.16 Single muon "OR" trigger efficiencies with respect to nseg3 and nseg1 offline qualities

6.3.7 Trigger and Offline Efficiencies

The Tag-and-Probe Method

In order to measure lepton reconstruction, identification, and trigger efficiencies, we use the "tag-and-probe" method. The idea is to select a sample of $Z/\gamma^* \to \ell^+\ell^-$ candidate events without imposing, e.g., a central track requirement on one of the leptons (the probe). This requires an independent measurement of the probe lepton momentum which can be used to reconstruct the dilepton invariant mass, $M_{\ell\ell}$. Requiring the $M_{\ell\ell}$ to be close to m_Z is normally sufficient to ensure that the event

6.3 Corrections to the Fully Simulated Monte Carlo Events 97

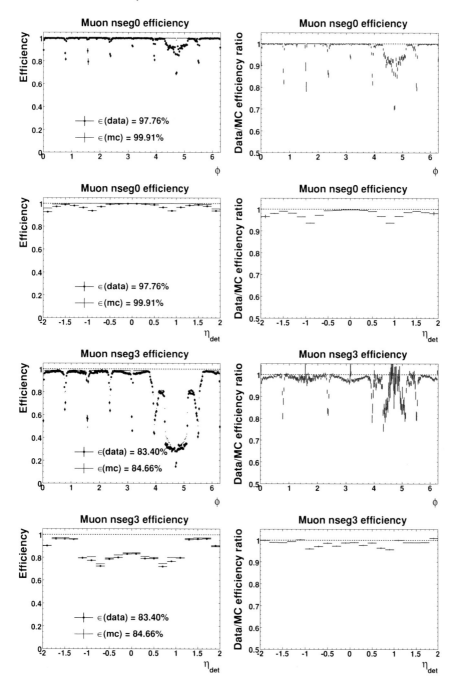

Fig. 6.17 Nseg $\geq$ 0, and nseg $\geq$ 3 local muon reconstruction (*left*) efficiency, and (*right*) data-vs-MC scale factor. The nseg $\geq$ 3 efficiency is with respect to a muon of at least nseg = 0 quality

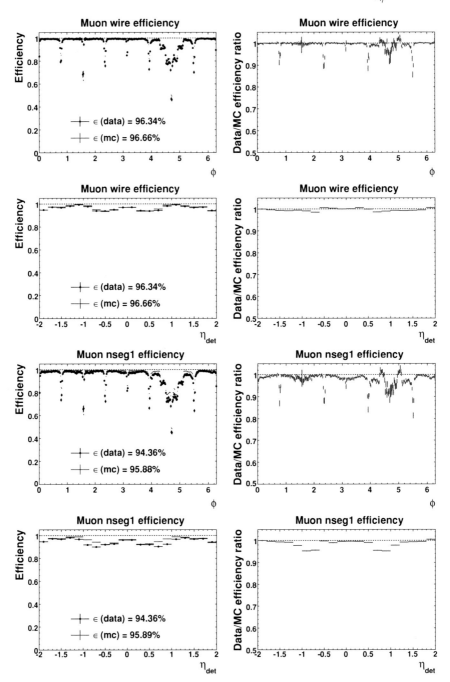

Fig. 6.18 Nseg1, and nseg0 (with wire hits) local muon reconstruction (*left*) efficiency, and (*right*) data-vs-MC scale factor. For both muon qualities, the efficiency is measured with respect to a muon of at least nseg = 0 quality

6.3 Corrections to the Fully Simulated Monte Carlo Events

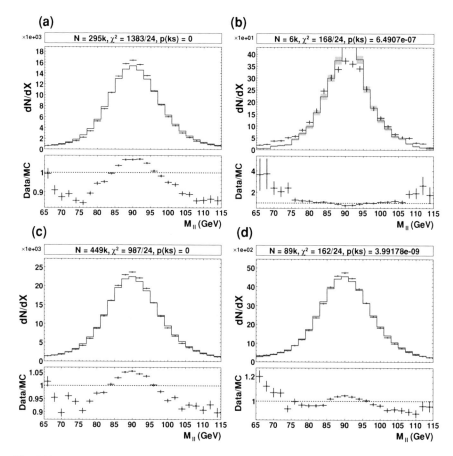

Fig. 6.19 Data-vs-MC comparison of the tag-and-probe invariant mass distributions separately for events where the probe passes/fails the studied requirement. The *black points* with error bars are the data, and the *red histogram* is the simulation. **a** Muon nseg0 (pass). **b** Muon nseg0 (fail). **c** Muon nseg3 (pass). **d** Muon nseg3 (fail)

sample is dominated by genuine $Z/\gamma^* \to \ell^+\ell^-$ decays. In order to eliminate bias, the tag lepton must be matched to an online trigger.

The efficiencies can be measured using exactly the same method in the MC, such that correction factors can be evaluated. The tag-and-probe method is prone to biases. For example, requiring the tag lepton to be matched to a central track biased the event sample towards periods of data taking with higher average tracking efficiency. However, these effects are at least partially cancelled when evaluating data-vs-MC efficiency correction factors.

100   6 Measurement of the Drell-Yan ϕ_η^* Distribution

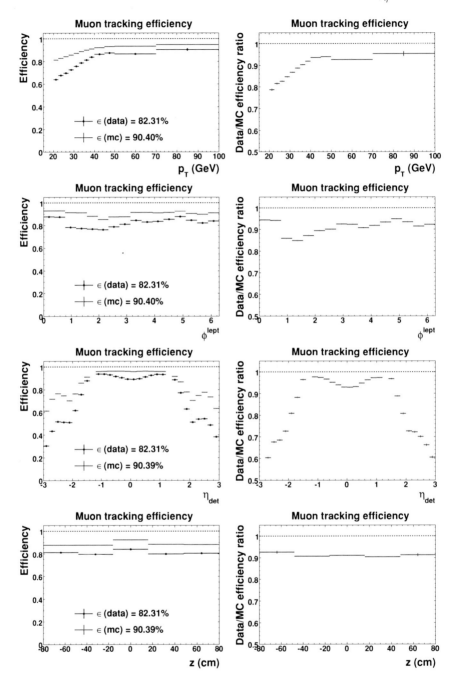

Fig. 6.20 Muon central track reconstruction (*left*) efficiency and (*right*) data-vs-MC scale factor

6.3 Corrections to the Fully Simulated Monte Carlo Events

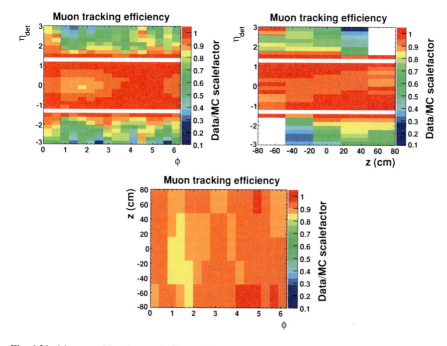

Fig. 6.21 Muon tracking data-vs-MC scale factors in various 2D projections

The detailed selection requirements for the tag-and-probe samples can be found in Ref. [7]. The different efficiency measurements are described in the following sub-sections. In all cases, we require a $Z/\gamma^* \rightarrow \ell^+\ell^-$ candidate with an invariant mass between 75 and 115 GeV, though the $M_{\ell\ell}$ distributions are presented over a wider range.

Single-EM Triggers

The efficiency for our offline electrons to have fired one of the single-EM triggers, is presented in Fig. 6.5, separately for the CC and EC. For CC electrons, the efficiency is simulated as function of ϕ_{mod} and E_T. For EC electrons, the efficiency is simulated as function of η_{det} and E_T.

Electron Reconstruction

Figures 6.6, 6.7 and 6.8 show the efficiencies for reconstructing EM clusters that have no trigger bias. A correction is applied to the MC in two dimensions (ϕ_{mod} and η_{det}) for CC electrons, and in one dimension (η_{det}) for EC electrons. Figure 6.9 shows the

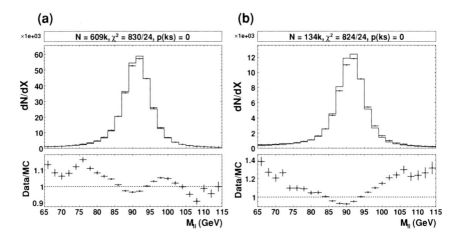

Fig. 6.22 Data-vs-MC comparison of the tag-and-probe invariant mass distributions separately for events where the probe passes/fails the studied requirement. The *black points* with error bars are the data, and the *red histogram* is the simulation. **a** Muon tracking (pass). **b** Muon tracking (fail)

$M_{\ell\ell}$ distribution of the electron plus track $Z/\gamma^* \to e^+e^-$ candidate events used to measure these efficiencies.

Electron Central Track Reconstruction

Figures 6.10 and 6.11 show the efficiencies to reconstruct a central track associated with an electron. A correction is applied to the MC in three dimensions (ϕ, η_{det}, and z_{dca}) for both CC and EC electrons. Figure 6.12 shows the $M_{\ell\ell}$ distribution of the $Z/\gamma^* \to e^+e^-$ candidate events that are used to measure the tracking efficiency.

Electron Quality Cuts

Figure 6.13 shows the efficiency for a CC EM cluster with a central track match to satisfy the higher level quality requirements. Figure 6.14 shows the equivalent for EC clusters. Corrections are applied to the MC in the same dimensions as for the cluster reconstruction efficiency. Figure 6.15 shows the $M_{\ell\ell}$ distribution of the $Z/\gamma^* \to e^+e^-$ candidate events that are used to measure these efficiencies.

Single-Muon Triggers

The efficiency for the selected offline muons to fire any of the single muon triggers is shown in Fig. 6.16, separately for nseg $\geq$ 1 (nseg1) and nseg = 3 (nseg3) local

6.3 Corrections to the Fully Simulated Monte Carlo Events

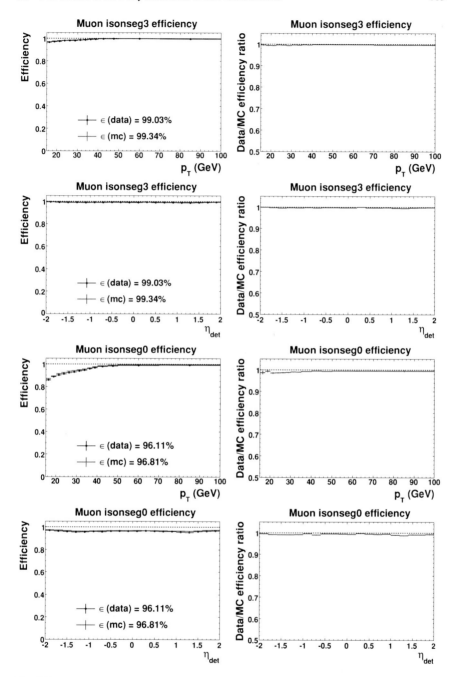

Fig. 6.23 Muon isolation (*left*) efficiencies, and (*right*) data-vs-MC scale factors

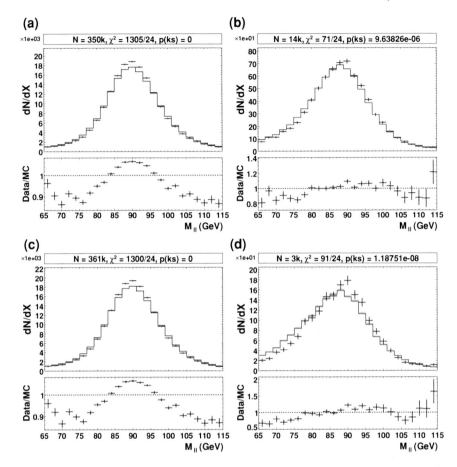

Fig. 6.24 Data-vs-MC comparison of the tag-and-probe invariant mass distributions separately for events where the probe passes/fails the studied requirement. The *black points* with error bars are the data, and the *red histogram* is the simulation. **a** Muon isonseg0 (pass). **b** Muon isonseg0 (fail). **c** Muon isonseg3 (pass). **d** Muon isonseg3 (fail)

muon qualities. Since the analysis trigger matching requirement is made for nseg3 muons, the nseg1 plots are only shown for interest of comparison. The efficiency is corrected in two dimensions (ϕ and η_{det}).

Local Muon Reconstruction

Figures 6.17 and 6.18 show the efficiencies to reconstruct local muons of nseg0, nseg1 and nseg3 qualities. The nseg1 and nseg3 efficiencies are determined with respect to a muon of nseg0 quality. In addition, Fig. 6.18 shows the efficiency for an nseg0 muon to have local muon hits (really only wire hits, since nseg0 muons cannot

6.3 Corrections to the Fully Simulated Monte Carlo Events

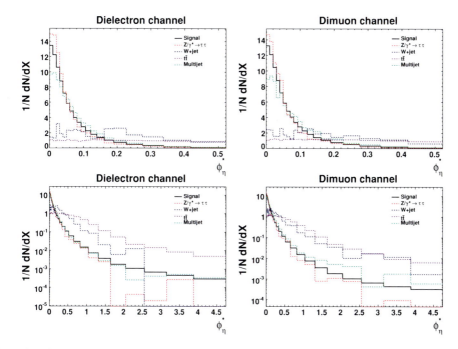

Fig. 6.25 Comparison of the shape of the ϕ^*_η distributions of the signal and background processes

Fig. 6.26 *Colour codes* used to distinguish the signal and background processes in this chapter

have scintillator hits), with respect to any nseg0 muon. The nseg0 wire efficiency plot is only included for interest, and is not used in the analysis. The MC is corrected by two-dimensional (ϕ and η_{det}) scale factors. Figure 6.19 shows the $M_{\ell\ell}$ distribution of the $Z/\gamma^* \to \mu^+\mu^-$ candidate events that are used to measure these efficiencies.

Muon Central Track Reconstruction

Figure 6.20 shows the tracking efficiency for muons, in bins of ϕ, η_{det}, and z_{dca}. All possible two-dimensional projections of the data-vs-MC scale factors are presented in Fig. 6.21. A correction is applied to the MC in three dimensions (ϕ, η_{det}, and z_{dca}).

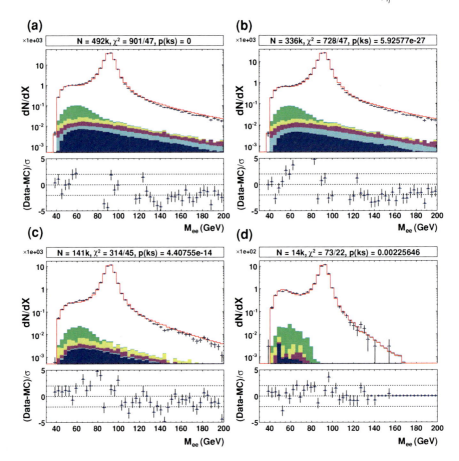

Fig. 6.27 Data-vs-MC comparison of the $M_{\ell\ell}$ distribution in the dielectron channel. **a** y-inclusive. **b** |y| < 1. **c** 1 ≤ |y| < 2. **d** |y| > 2

In the "bottom hole" region, $|\eta_{det}| < 1.1$, $4.2 < \phi < 5.1$, the muon system is not sufficiently instrumented to be able to measure the p_T of muon candidates. Since no significant variation in the electron tracking efficiency scale factor is observed in this region, we choose a scale factor of 0.95, which is close to that of the surrounding regions. Figure 6.22 shows the $M_{\ell\ell}$ distribution of the $Z/\gamma^* \to \mu^+\mu^-$ candidate events that are used to measure the tracking efficiencies.

Muon Isolation

Figure 6.23 shows the efficiencies for the different muon isolation requirements used in this analysis. Since the efficiencies are high, and agree reasonably well between data and MC, no corrections are applied to the MC. Figure 6.24 shows the $M_{\ell\ell}$

6.3 Corrections to the Fully Simulated Monte Carlo Events

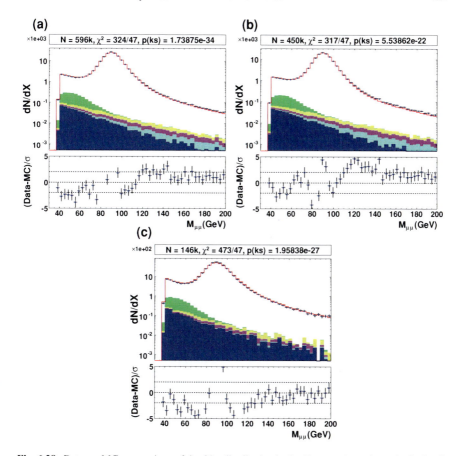

Fig. 6.28 Data-vs-MC comparison of the $M_{\ell\ell}$ distribution in the dimuon channel. **a** y-inclusive. **b** |y| < 1. **c** $1 \leq |y| < 2$

distribution of the $Z/\gamma^* \to \mu^+\mu^-$ candidate event that are used to measure these efficiencies.

Additional Corrections to the Efficiencies

The following corrections are applied to the single-muon trigger efficiencies in order to improve agreement in the relevant distributions.

- Scale by a factor of 1.2 if $|\eta_{\text{det}}| > 1.1$ and $|\eta_{\text{det}}| < 1.5$.
- Scale by a factor of 0.85 if $|\eta_{\text{det}}| < 0.5$.
- Scale by a factor of 1.2 if $\phi_{\text{mod}}^{(8)} > 0.47$, where $\phi_{\text{mod}}^{(8)}$ is equivalent to ϕ_{mod} except with 8 modules rather than 32.

Fig. 6.29 Data-vs-MC comparison of the α_{aco} distribution in our selected dimuon events

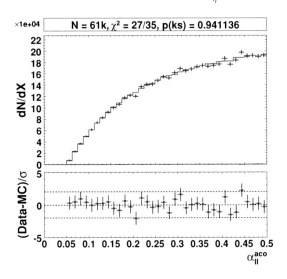

In addition, the CC electron reconstruction efficiency scale factor is increased by a factor of 1.2 for CC electrons with $\phi_{\text{mod}} > 0.47$.

6.4 Backgrounds

Figure 6.25 shows a comparison of the shape of the ϕ_η^* distributions for the various background processes considered (except cosmic ray muons). The distribution of ϕ_η^* is significantly broader in W (+jet) events than in the other backgrounds or the $Z/\gamma^* \to \ell^+\ell^-$ signal. For future reference, Fig. 6.26 shows the legend used to distinguish the different contributions in the data-MC comparison plots in the rest of this chapter.

Figures 6.27 and 6.28 show for the dielectron and dimuon channels respectively that the $M_{\ell\ell}$ distribution in data is adequately described by the sum of signal and predicted backgrounds. The background sources are considered as follows:

- **Physics backgrounds** The processes, $Z/\gamma^* \to \tau^+\tau^-$, $WW \to \ell^+\nu\ell^-\bar{\nu}$, and $t\bar{t} \to ll\nu\nu b\bar{b}$ are simulated in the same way as the signal MC.
- **W (+jet) events (dielectron only)** Additional jets in $W \to e\nu$ events can fake electrons. MC events are produced, and normalised in the same way as the signal. These events have a very low probability to pass our event selection, but carry large weights, due to the relatively large cross section. The contribution is $\sim 0.05\%$, which we consider small enough to be ignored in the rest of the analysis, which avoids "nuisance" events with very large weight.
- **QCD Multijets** In the dielectron channel, these contributions are from QCD multijet events where two jets are mis-identified as electrons. As there are no sufficiently

6.4 Backgrounds

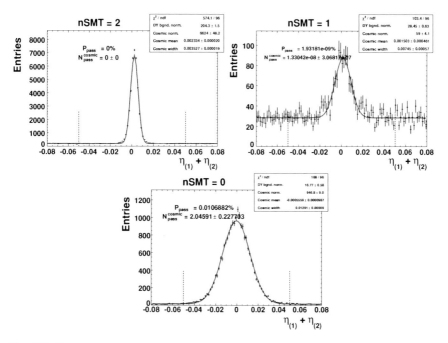

Fig. 6.30 The $\eta_{(1)} + \eta_{(2)}$ distribution in the dimuon dataset, for $n_{\text{SMT}} = 0, 1, 2$, before cutting on this variable. The *vertical lines* represent the cut window. The distribution is fitted with a Gaussian function, which parameterises the width and height of the cosmic peak, on top of a linear function to describe the Drell-Yan signal

accurate tools to generate MC multi-jet events, and the efficiency to select such events would be extremely low, data is used to estimate the shape of the QCD background. Compared to our signal selection, all higher level quality cuts are relaxed, and an inverted isolation requirement: $0.08 < \mathcal{I}_{\text{cal}}^{\text{EM}} < 0.15$, removes any remaining signal. In addition, we require $\not{E}_T < 15\,\text{GeV}$, in order to suppress W +jet events, which are simulated using MC. A maximum likelihood fit to the invariant mass distribution determines the relative normalisation of the MC (signal $+ Z/\gamma^* \to \tau^+\tau^- + W$) and the QCD template. This fit is performed separately for each analysis rapidity bin. In the dimuon channel, muonic decays, of light mesons (π, K) from QCD multi-jet events, or heavy quarks in $b\bar{b}$ and $c\bar{c}$ events, can mimic the dimuon signal, but are suppressed by the muon isolation cuts. The fraction on QCD events, within the selected event sample, is estimated using same sign events to be approximately 0.1 %. A sample of QCD events is selected from data, by inverting the isolation requirements ($\mathcal{I}_{\text{trk}}/p_T > 0.03$ and $\mathcal{I}_{\text{cal}}/p_T > 0.1$), relaxing the opposite sign requirement, dropping the trigger matching requirement, and allowing events to also fire dimuon triggers. This sample is normalised to the fraction determined above.

110   6 Measurement of the Drell-Yan ϕ_η^* Distribution

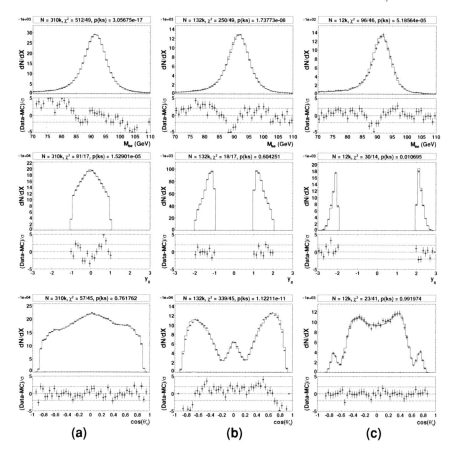

Fig. 6.31 Data-vs-MC comparisons of the invariant mass distribution in the dielectron channel. **a** |y| < 1. **b** 1 ≤ |y| < 2. **c** |y| > 2

- **Cosmic ray muons (dimuon only)** These are a potential danger, since they have zero ϕ_η^*. The pseudo-acolinearity variable, defined as $\alpha_{\text{aco}} = |\eta_1 + \eta_2| + ||\phi_1 - \phi_2| - \pi|$, discriminates against cosmic ray muons, which have $\alpha_{\text{aco}} \sim 0$. Figure 6.29 shows that there is no excess of data in this region, and contamination from cosmics is claimed to be negligible. Our selection cuts exclude the region $|\eta_{(1)} + \eta_{(2)}| < 0.05$. Figure 6.30 shows that a prominent cosmic peak is observed in the $\eta_{(1)} + \eta_{(2)}$ distributions before applying this cut. A quantitative estimate of the cosmic contamination can be obtained by fitting the $\eta_{(1)} + \eta_{(2)}$ distribution with a Gaussian, which parameterises the width and height of the cosmic peak, and a linear function that describes the Drell-Yan signal (or in this study, background). Events are categorised based on the number of muons (0, 1 or 2) which have at least one hit in the SMT, n_{SMT}. The estimated cosmic contamination is negligible for the $n_{\text{SMT}} = 1, 2$ categories, and 2.0 ± 0.2 events for the $n_{\text{SMT}} = 0$ category.

6.4 Backgrounds

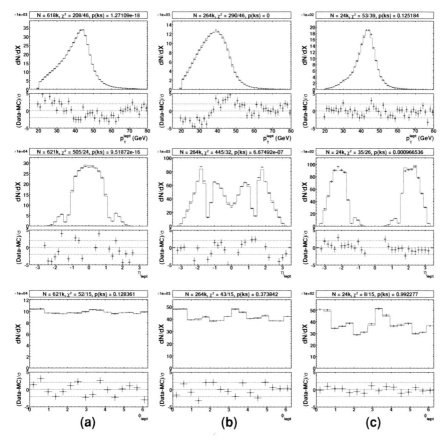

Fig. 6.32 Data-vs-MC comparisons of the lepton p_T distribution in the dielectron channel. **a** |y| < 1. **b** 1 ≤ |y| < 2. **c** |y| > 2

Since the purity of the first bin in ϕ^*_η is high (above 95%), all cosmic ray muon events would fall into this bin, in which the estimated contamination of ≈2 events can be compared to around 50k events in data (Figs. 6.31, 6.32, 6.33, 6.34, 6.35, 6.36, 6.37, 6.38, 6.39).

6.5 Comparison of Data with Simulation

In this section, the data are compared to the corrected signal plus background simulation in a large number of distributions. For each bin, the content is divided by the width (dN/dX). The level of agreement is considered sufficient for the purposes of this analysis. In most cases, the lower halves of the plots show (Data-MC)/σ which

112   6 Measurement of the Drell-Yan ϕ^*_η Distribution

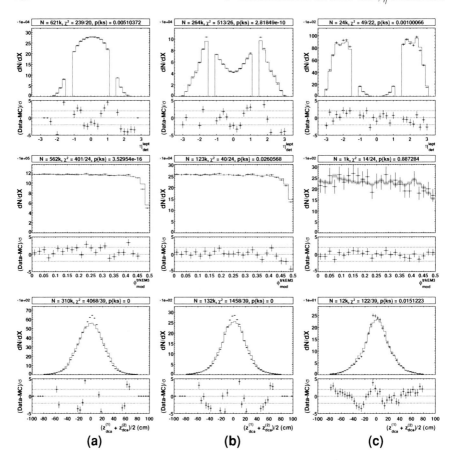

Fig. 6.33 Data-vs-MC comparisons of the lepton η_{det} distribution in the dielectron channel. **a** |y| < 1. **b** 1 ≤ |y| < 2. **c** |y| > 2

is the difference between the data and MC, divided by the statistical uncertainty on the difference (data and MC uncertainties added in quadrature). In some cases, the ratio of data to MC is considered to be more appropriate and is shown instead.

6.5 Comparison of Data with Simulation

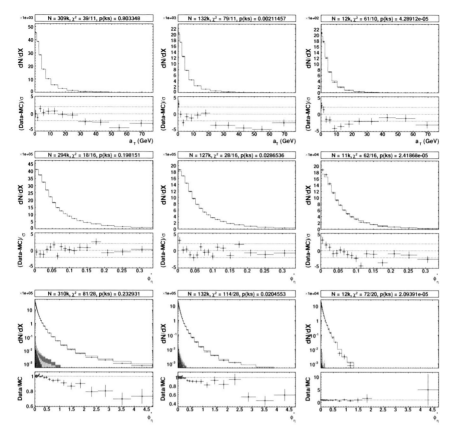

Fig. 6.34 Data-vs-MC comparisons of the a_T abd ϕ_η^* distributions in the dielectron channel. **a** |y| < 1. **b** 1 ≤ |y| < 2. **c** |y| > 2

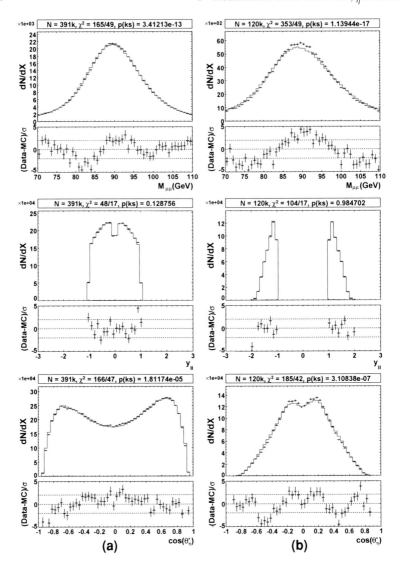

Fig. 6.35 Data-vs-MC comparisons of the invariant mass distribution in the dimuon channel. **a** |y| < 1. **b** 1 ≤ |y| < 2

6.5 Comparison of Data with Simulation 115

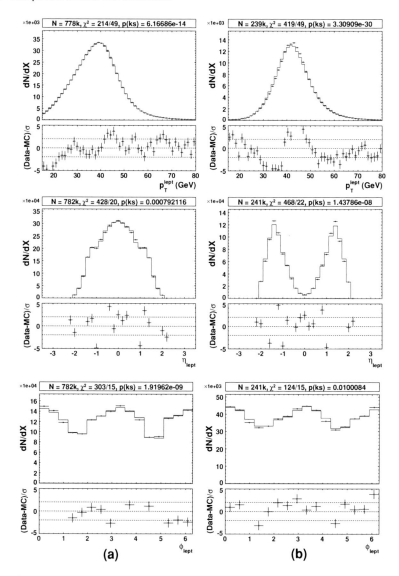

Fig. 6.36 Data-vs-MC comparisons of the lepton p_T distribution in the dimuon channel. **a** |y| < 1. **b** 1 ≤ |y| < 2

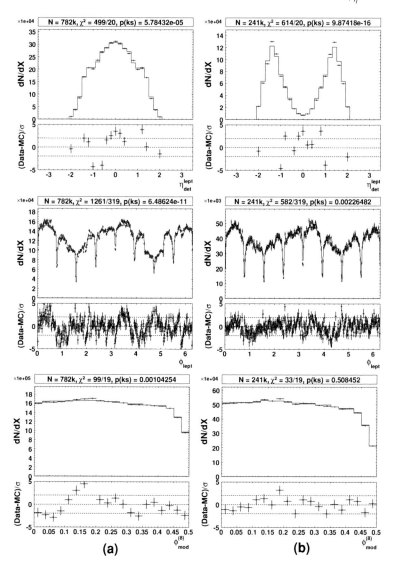

Fig. 6.37 Data-vs-MC comparisons of the lepton η_{det} distribution in the dimuon channel. **a** $|y| < 1$. **b** $1 \leq |y| < 2$

6.5 Comparison of Data with Simulation

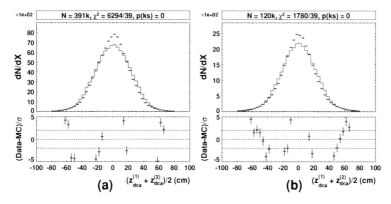

Fig. 6.38 Data-vs-MC comparisons of the z_{PV} distribution in the dimuon channel. **a** $|y| < 1$. **b** $1 \leq |y| < 2$

Table 6.1 Comparison of g_2^{KF} and g_2 fits for different ϕ-gap and rapidity requirements, separately for the dimuon and dielectron channels

Category	g_2^{KF} (GeV^2)	Significance (σ)	g_2 (GeV^2)	Significance (σ)
		$|y| < 1$		
ee, 2 MUfid	0.696 ± 0.013	Reference	0.649 ± 0.013	Reference
ee, 1 MUfid	0.690 ± 0.033	-0.2	0.698 ± 0.033	1.4
ee, 0 MUfid	0.724 ± 0.039	0.7	0.681 ± 0.039	0.8
ee, 2 EMfid	0.720 ± 0.013	Reference	0.673 ± 0.013	Reference
ee, 1 EMfid	0.663 ± 0.024	-2.0	0.632 ± 0.024	-1.5
ee, 0 EMfid	0.601 ± 0.061	-1.9	0.527 ± 0.063	-2.3
$\mu\mu$, 2 MUfid	0.707 ± 0.012	Reference	0.658 ± 0.012	Reference
$\mu\mu$, 1 MUfid	0.758 ± 0.033	1.5	0.763 ± 0.033	3.0
$\mu\mu$, 0 MUfid	0.802 ± 0.042	2.2	0.757 ± 0.042	2.3
$\mu\mu$, 2 EMfid	0.720 ± 0.012	Reference	0.671 ± 0.012	Reference
$\mu\mu$, 1 EMfid	0.762 ± 0.023	1.6	0.726 ± 0.023	2.1
$\mu\mu$, 0 EMfid	0.792 ± 0.061	1.2	0.736 ± 0.063	1.0
		$1 \leq |y| < 2$		
ee, 2 MUfid	0.618 ± 0.018	Reference	0.575 ± 0.018	Reference
ee, 1 MUfid	0.730 ± 0.048	2.2	0.737 ± 0.048	3.2
ee, 0 MUfid	0.661 ± 0.058	0.7	0.607 ± 0.061	0.5
ee, 2 EMfid	0.629 ± 0.017	Reference	0.588 ± 0.017	Reference
ee, 1 EMfid	0.663 ± 0.039	0.8	0.621 ± 0.040	0.8
ee, 0 EMfid	0.740 ± 0.100	1.1	0.740 ± 0.100	1.5
$\mu\mu$, 2 MUfid	0.658 ± 0.020	Reference	0.612 ± 0.018	Reference
$\mu\mu$, 1 MUfid	0.644 ± 0.061	-0.2	0.655 ± 0.061	0.7
$\mu\mu$, 0 MUfid	0.840 ± 0.064	2.7	0.813 ± 0.064	3.0
$\mu\mu$, 2 EMfid	0.665 ± 0.019	Reference	0.623 ± 0.018	Reference
$\mu\mu$, 1 EMfid	0.766 ± 0.050	1.9	0.729 ± 0.051	2.0
$\mu\mu$, 0 EMfid	0.740 ± 0.100	0.7	0.740 ± 0.100	1.2

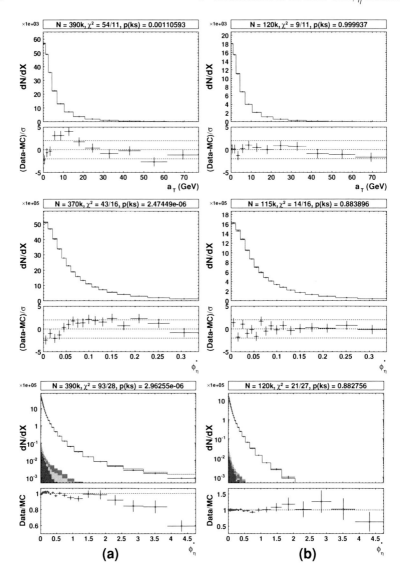

Fig. 6.39 Data-vs-MC comparisons of the a_T and ϕ_η^* distributions in the dimuon channel. **a** |y| < 1. **b** 1 ≤ |y| < 2

6.5 Comparison of Data with Simulation 119

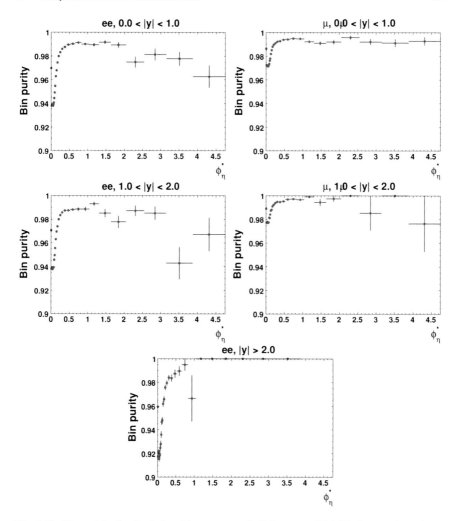

Fig. 6.40 Bin purities for the (*left* and *bottom* central) dielectron and (*right*) dimuon channels

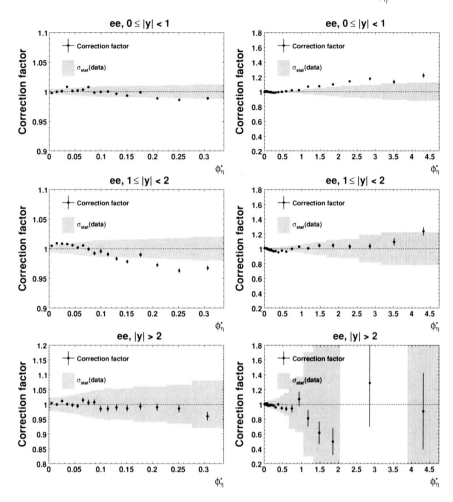

Fig. 6.41 Bin-by-bin correction factors in the dielectron channel. The *yellow bands* represent the size of the data statistical uncertainties

6.5 Comparison of Data with Simulation

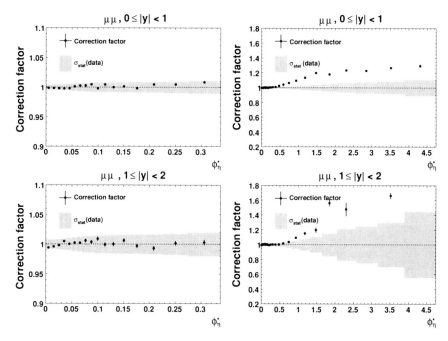

Fig. 6.42 Bin-by-bin correction factors in the dimuon channel. The *yellow bands* represent the size of the data statistical uncertainties

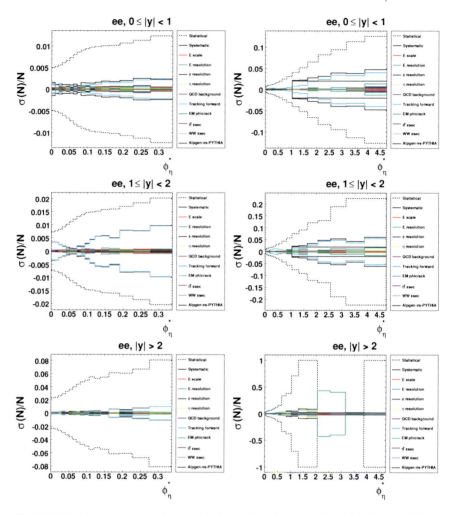

Fig. 6.43 Experimental systematic uncertainties for the dielectron channel in bins of ϕ^*. The *left-hand plots* show the region of low ϕ^* and the *right-hand plots* show the full range in ϕ^*

6.5 Comparison of Data with Simulation

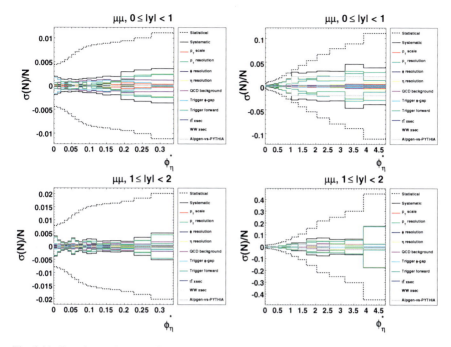

Fig. 6.44 Experimental systematic uncertainties for the dimuon channel in bins of ϕ^*. The *left-hand plots* show the region of low ϕ^* and the *right-hand plots* show the full range in ϕ^*

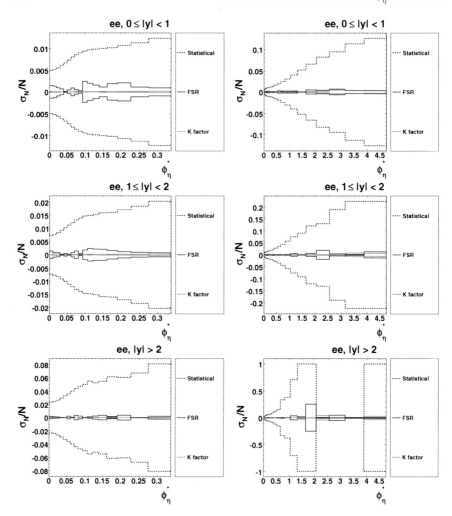

Fig. 6.45 Systematic uncertainties due to physics input bias for the dielectron channel. The *left-hand plots* show the region of low ϕ^* and the *right-hand plots* show the full range in ϕ^*

6.5 Comparison of Data with Simulation

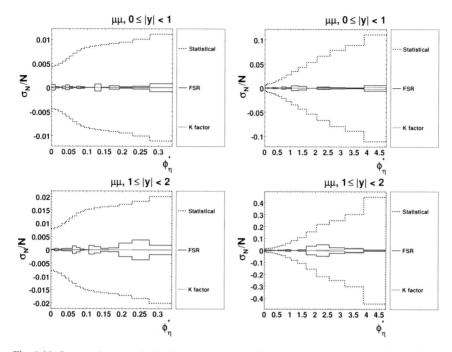

Fig. 6.46 Systematic uncertainties due to physics input bias for the dimuon channel. The *left-hand plots* show the region of low ϕ^* and the *right-hand plots* show the full range in ϕ^*

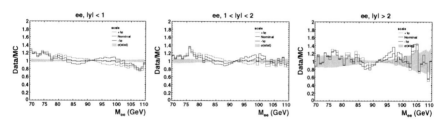

Fig. 6.47 Energy scale systematic uncertainties for the dielectron channel: data/MC ratios in the $M_{\ell\ell}$ distribution for the positive and negative shifts

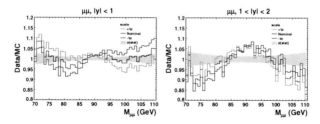

Fig. 6.48 Momentum scale systematic uncertainties for the dimuon channel: data/MC ratios in the $M_{\ell\ell}$ distribution for the positive and negative shifts

126 6 Measurement of the Drell-Yan ϕ_η^* Distribution

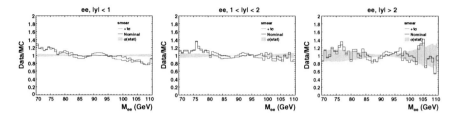

Fig. 6.49 Energy resolution systematic uncertainties for the dielectron channel: data/MC ratios in the $M_{\ell\ell}$ distribution for the nominal and additional smearing

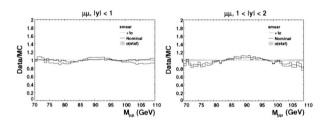

Fig. 6.50 Track p_T resolution systematic uncertainties for the dimuon channel: data/MC ratios in the $M_{\ell\ell}$ distribution for the nominal and additional smearing

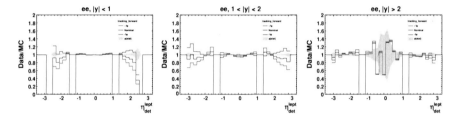

Fig. 6.51 Tracking forward (efficiency for electrons) systematic uncertainties for the dielectron channel: data/MC ratio in the η_{det} distribution for the positive and negative variations

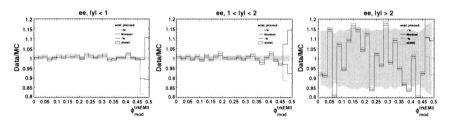

Fig. 6.52 EM phicrack (cluster reconstruction efficiency for electrons) systematic uncertainties for the dielectron channel: data/MC ratio in the ϕ_{mod} distribution for the positive and negative variations

6.5 Comparison of Data with Simulation 127

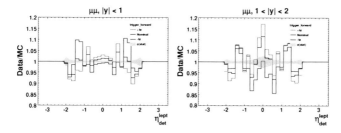

Fig. 6.53 Muon trigger efficiency systematic uncertainties for the dimuon channel: data/MC ratios for the η_{det} distribution for the positive and negative variations

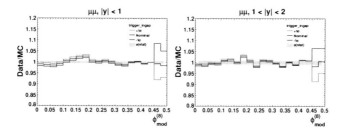

Fig. 6.54 Muon trigger efficiency systematic uncertainties for the dimuon channel: data/MC ratios for the ϕ_{mod} distribution for the positive and negative variations

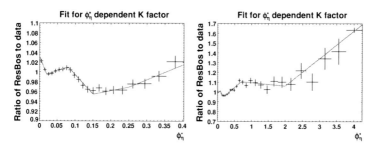

Fig. 6.55 Ratio of RESBOS to the unfolded dielectron data, to be used in determining sensitivity to physics input biases

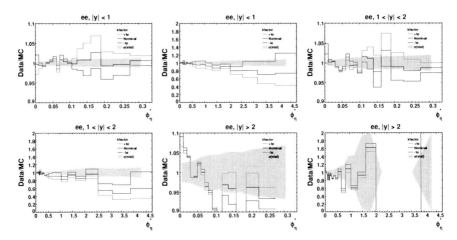

Fig. 6.56 Systematic uncertainties due to physics input bias: the ratio of the RESBOS to the data for dielectrons. The data/MC ratios are shown (at folded level) when (i) the MC is scaled by this ratio, and (ii) the MC is scaled by the inverse of this ratio. The plots on the *left* correspond are over a restricted ϕ_η^* range, and those on the *right* are over the full range

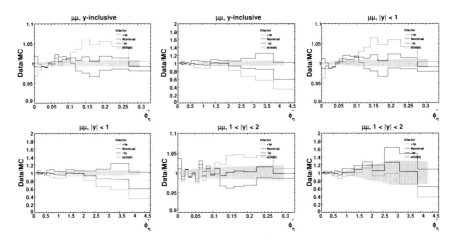

Fig. 6.57 Systematic uncertainties due to physics input bias: the ratio of the RESBOS to the data for dimuons. The data/MC ratios are shown (at folded level) when (i) the MC is scaled by this ratio, and (ii) the MC is scaled by the inverse of this ratio. The plots on the *left* correspond are over a restricted ϕ_η^* range, and those on the *right* are over the full range

6.5 Comparison of Data with Simulation

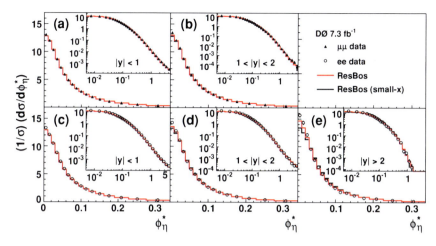

Fig. 6.58 Comparison of the measured ϕ_η^* distribution with predictions from RESBOS. The *black* and *red histograms* correspond to predictions from RESBOS with and without the small-x broadening effect

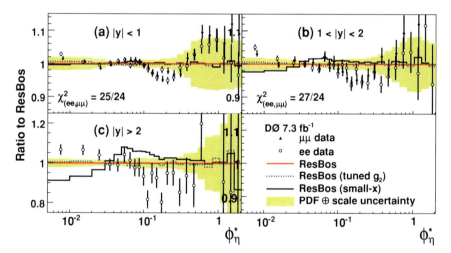

Fig. 6.59 The ratio to the nominal RESBOS prediction of; the dimuon and dielectron data, and alternative RESBOS predictions with (*blue*) the value of g_2 that best describes the data, and (*black*) the small-x broadening effect

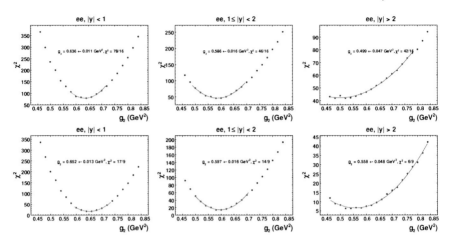

Fig. 6.60 Fits for g_2 using the unfolded ϕ_η^* distributions. The *bottom row* restricts the fit to the first 10 bins in ϕ_η^*

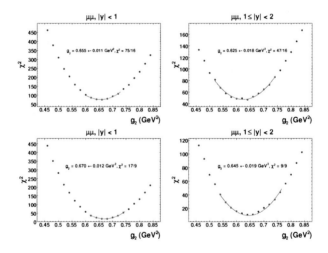

Fig. 6.61 Fits for g_2 using the unfolded ϕ_η^* distributions. The *bottom row* restricts the fit to the first 10 bins in ϕ_η^*

6.5 Comparison of Data with Simulation

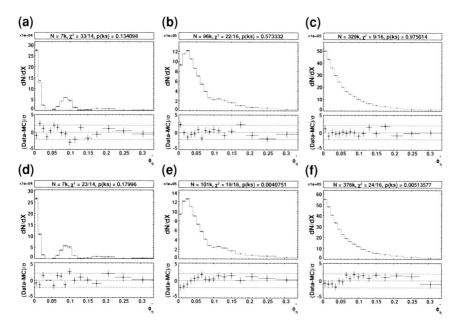

Fig. 6.62 Data-vs-MC comparison of the ϕ_η^* distributions in events with 0, 1, and 2 leptons within the CC electron fiducial acceptance. **a** Dielectron 0 CCfid. **b** Dielectron 1 CCfid. **c** Dielectron 2 CCfid. **d** Dimuon 0 CCfid. **e** Dimuon 1 CCfid. **f** Dimuon 2 CCfid

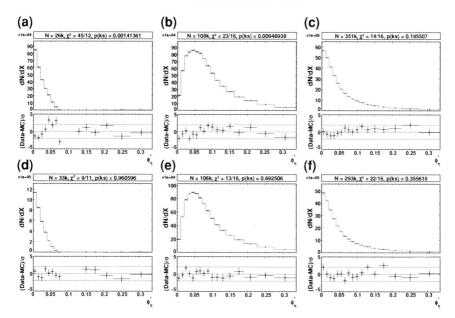

Fig. 6.63 Data-vs-MC comparison of the ϕ_η^* distributions in events with 0, 1, and 2 leptons within the muon fiducial acceptance. **a** Dimuon 0 MUfid. **b** Dimuon 1 MUfid. **c** Dimuon 2 MUfid. **d** Dielectron 0 MUfid. **e** Dielectron 1 MUfid. **f** Dielectron 2 MUfid

132 6 Measurement of the Drell-Yan ϕ^*_η Distribution

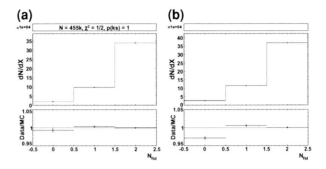

Fig. 6.64 Data-vs-MC comparison of the number of fiducial leptons. **a** Dielectron. **b** Dimuon

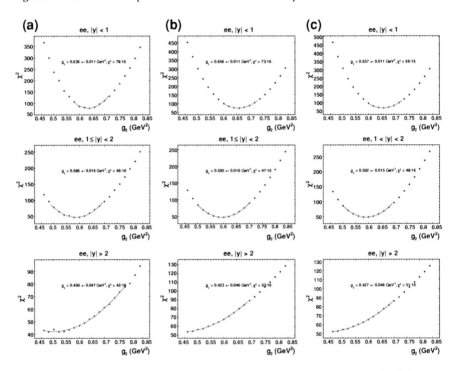

Fig. 6.65 Fits for g_2 in the dielectron channel, performed at the unfolded level with (*left*) RES-BOS/PHOTOS and (*middle*) re-weighted PYTHIA, and (*right*) at the un-corrected detector level. **a** unfolded-level (ResBos). **b** unfolded-level (pythia). **c** detector-level (pythia)

6.6 Unfolding   133

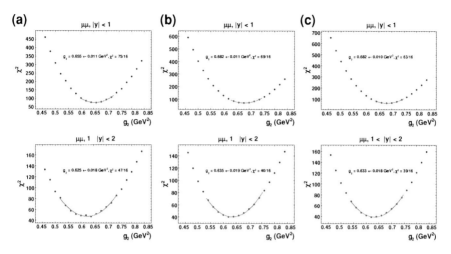

Fig. 6.66 Fits for g_2 in the dimuon channel, performed at the unfolded level with (*left*) RES-BOS/PHOTOS and (*middle*) re-weighted PYTHIA, and (*right*) at the un-corrected detector level. **a** unfolded-level (ResBos). **b** unfolded-level (pythia). **c** detector-level (pythia)

6.6 Unfolding

6.6.1 Binning in ϕ_η^*

We use 40 bins, with the width of bin i (where the first bin corresponds to $i = 1$) given by $0.01 + 5 \times 10^{-8} \times (i-1)^5$. For $|y| > 2$, only the first 30 bins are considered, since the expected yield above this is very small. The last (30th or 40th depending on the $|y|$ range) bin is combined with the overflow. We shall see later that the total uncertainties are dominated by statistics for all bins. The bins are therefore considered to be narrow enough.

6.6.2 Bin-by-Bin Unfolding

Once we are satisfied that the fully simulated MC event sample describes the data with sufficient accuracy, we can use it to correct the data back to particle level. Since the experimental resolution in ϕ_η^* is very good, which is apparent in the bin purities[2] shown in Fig. 6.40, a simple bin-by-bin unfolding procedure is reasonable. Figures 6.41 and 6.42 show, for the dielectron and dimuon channels respectively, the

---

[2] The bin purity is defined as the fraction of events in each bin at generator level, which are also in the same bin at detector level.

bin-by-bin correction factors (data points with error bars representing the MC statistical uncertainty). The yellow bands show the size of the data statistical uncertainties.

6.7 Systematic Uncertainties

Figures 6.43 and 6.44 show, for the dielectron and dimuon channels respectively, the considered systematic uncertainties for each bin in ϕ^*, compared to the statistical uncertainty. The left-hand plots show the region of low ϕ^* and the right-hand plots show the full range in ϕ^*. Additional sources of systematic uncertainty due to physics input biases are shown in Figs. 6.45 and 6.46. Each source of systematic uncertainty is described in more detail here:

- **Energy and momentum scales** The electron (muon) energy (momentum) scales in MC are shifted by $\pm 0.3\%$, which is larger than the $\sim 0.1\%$ uncertainty determined for electrons in Chap. 4. This reflects the fact that we are using very different electron quality cuts than those used by most DØ analyses (and the studies of Chap. 4) that include shower shape requirements. Figures 6.47 and 6.48 show, for the dielectron and dimuon channels respectively, the data/MC ratios in the $M_{\ell\ell}$ distribution for the positive and negative shifts.
- **Energy and momentum resolution** An additional smearing of width 0.02 (in $\Delta E/E$) is applied to electron energies in MC. For the dimuons, an additional smearing in $(1/p_T)$ of width $0.001\,\mathrm{GeV}^{-1}$ is treated as a systematic uncertainty. Figures 6.49 and 6.50 show, for the dielectron and dimuon channels respectively, the data/MC ratios in the $M_{\ell\ell}$ distribution for these variation, compared to the default resolution.
- **Track ϕ and η resolution** The ϕ and η smearing parameters are scaled up and down by a factor of 1.5, which is considered to be a reasonable estimate of the uncertainty.
- **Tracking forward (efficiency for electrons)** For EC electrons, the tracking efficiency is multiplied by the following factor:

$$1 + (0.6/1.5) \times (|\eta_{\mathrm{det}}| - 1.5)$$

The form and magnitude of this factor is chosen such that the data-vs-MC discrepancy in the η_{det} distribution is adequately covered. Figure 6.51 shows the data/MC ratio in the η_{det} distribution for the positive and negative variations.
- **EM phicrack (cluster reconstruction efficiency for electrons)** A $\pm 10\%$ shift in the efficiency for electrons in the CC ϕ-gaps ($\phi_{\mathrm{mod}} > 0.45$) is taken as a systematic uncertainty. Figure 6.52 shows the data/MC ratio in the ϕ_{mod} distribution for the positive and negative variations.
- **Muon trigger efficiency** The absolute trigger efficiency is unimportant, but the analysis is sensitive to the dependence on η_{det} and ϕ_{mod}. We consider a $\pm 10\%$ shift in the efficiency for (a) muons in the forward region, and (b) muons in the

6.7 Systematic Uncertainties

octant gaps ($\phi_{mod} > 0.45$), as a reasonable estimate of the uncertainty in these dependencies. Figure 6.53 and 6.54 show the data/MC ratios for η_{det} and ϕ_{mod} respectively, for the corresponding efficiency shifts.
- **Muon reconstruction and identification efficiency** A $\pm 10\%$ shift in the efficiency for (a) muons in the forward region, and (b) muons in the octant gaps, as a reasonable estimate of the uncertainty in these dependencies.
- **Physics input bias** Figure 6.55 shows the ratio of RESBOS to the unfolded data (weighted mean of dielectron and dimuon channels); this is used in determining the sensitivity of the unfolding correction to imperfections in the MC input physics distribution. Figures 6.56 and 6.57 show, for the dielectrons and dimuons respectively, the ratio of RESBOS to the data. The data/MC ratios are shown (at uncorrected level) when (i) the MC is scaled by this ratio, and (ii) the MC is scaled by the inverse of this ratio.
- **Final state radiation (FSR)** The parameter Δ_{QED} is defined as the difference in invariant mass between the propagator level Z/γ^* and the particle level dilepton (after FSR). Events with $\Delta_{QED} > 0.5\,\text{GeV}$ are scaled up/down by a factor of two.
- **Dependence on choice of event generator** Differences in the ϕ_η^* dependence of the isolation efficiency could arise between PYTHIA and e.g. ALPGEN due to the different parton level multiplicities. Unfortunately, there are insufficient ALPGEN statistics available to be able to precisely compare the detector corrections. A simple fast simulation of the detector effects (resolutions and efficiencies) shows a disagreement of approximately 2% between PYTHIA and ALPGEN at large ϕ_η^*. This is only observed for the dielectron channel, and $|y| < 1$, where a 2% systematic uncertainty is assigned for $\phi_\eta^* > 1$.

6.8 Results

In this section, we present the measured ϕ_η^* distributions corrected for experimental acceptance and resolution, and compare these with theoretical predictions. Tables of the corrected ϕ_η^* distributions have been published in [8].

6.8.1 Theoretical Predictions

Event samples are generated using RESBOS [1], interfaced with PHOTOS [2] to simulate the effect of QED final state radiation (FSR). Rather than correct the data for FSR, we choose to compare the data to a prediction with the effect of FSR simulated. From now on RESBOS refers to RESBOS + PHOTOS (where not stated explicitly). The following samples are generated:

- Default RESBOS settings (CTEQ 6.6 PDFs [9], and BLNY form factor [10] with $g_1 = 0.21$, $g_2 = 0.68\,\text{GeV}^2$, $g_3 = -0.6$).
- Default settings, except with g_2 ranging from 0.46 to 0.84 GeV^2 in increments of 0.02 GeV^2.

Table 6.2 Comparison of the g_2 fits for data which passes and fails standard data quality requirements

Category	$|y| < 1$		$1 \leq |y| < 2$	
ee, Pass dq	0.657 ± 0.012	Reference	0.596 ± 0.016	Reference
ee, Fail dq	0.673 ± 0.028	+0.6	0.581 ± 0.042	−0.3
ee, nSMT = 2	0.651 ± 0.012	Reference	0.586 ± 0.016	Reference
ee, nSMT = 1	0.686 ± 0.031	+1.1	0.675 ± 0.053	+1.6
ee, nSMT = 0	0.748 ± 0.052	+1.8	0.560 ± 0.100	−0.3
ee, ilum < 50	0.637 ± 0.024	Reference	0.596 ± 0.032	Reference
ee, 50 < ilum < 150	0.662 ± 0.015	+0.9	0.598 ± 0.019	+0.1
ee, ilum > 150	0.685 ± 0.025	+1.4	0.589 ± 0.037	−0.1
ee, $|z| < 40$	0.642 ± 0.013	Reference	0.590 ± 0.018	Reference
ee, $|z| > 40$	0.697 ± 0.018	+2.4	0.607 ± 0.028	+0.5
ee, sptp	0.653 ± 0.020	Reference	0.614 ± 0.034	Reference
ee, sntp	0.642 ± 0.021	−0.4	0.600 ± 0.029	−0.3
ee, sptn	0.653 ± 0.020	−0.0	0.571 ± 0.029	−1.0
ee, sntn	0.696 ± 0.020	+1.5	0.605 ± 0.030	−0.2
$\mu\mu$, Pass dq	0.677 ± 0.011	Reference	0.632 ± 0.019	Reference
$\mu\mu$, Fail dq	0.719 ± 0.027	+1.4	0.647 ± 0.044	+0.3
$\mu\mu$, nSMT = 2	0.680 ± 0.011	Reference	0.629 ± 0.019	Reference
$\mu\mu$, nSMT = 1	0.700 ± 0.028	+0.7	0.664 ± 0.061	+0.6
$\mu\mu$, nSMT = 0	0.694 ± 0.050	+0.3	0.693 ± 0.100	+0.6
$\mu\mu$, ilum < 50	0.665 ± 0.020	Reference	0.643 ± 0.038	Reference
$\mu\mu$, 50 < ilum < 150	0.686 ± 0.014	+0.8	0.645 ± 0.024	+0.1
$\mu\mu$, ilum > 150	0.672 ± 0.024	+0.2	0.613 ± 0.041	−0.5
$\mu\mu$, $|z| < 40$	0.681 ± 0.013	Reference	0.630 ± 0.020	Reference
$\mu\mu$, $|z| > 40$	0.689 ± 0.018	+0.4	0.648 ± 0.031	+0.5
$\mu\mu$, sptp	0.663 ± 0.019	Reference	0.665 ± 0.035	Reference
$\mu\mu$, sntp	0.686 ± 0.018	+0.9	0.644 ± 0.033	−0.4
$\mu\mu$, sptn	0.711 ± 0.018	+1.8	0.637 ± 0.035	−0.6
$\mu\mu$, sntn	0.678 ± 0.020	+0.5	0.613 ± 0.033	−1.1

The different categories are described in the text. The units of $\mathcal{L}_{\text{inst}}$ are $\times 10^{30}$ cm^{-2} s^{-1}

- Default settings, except with 44 CTEQ 6.6 PDF error sets.
- Default settings, except with renormalisation and factorisation scales varied by a factor of two.
- Default settings, except with an additional x dependence (so-called small-x broadening) of the NP form factor. This includes an additional Gaussian form factor of width,

$$\rho(x) = c_0 \left(\sqrt{\frac{1}{x^2} + \frac{1}{x_0^2}} - \frac{1}{x_0} \right),$$

where $x_0 = 0.005$ and $c_0 = 0.013$ [11].

In all cases, around 10 M *weighted* events are generated for both up- and down-type quark annihilation (these are generated separately in the CP version of RESBOS). All

6.8 Results

available grid files are truncated at $p_T < 380\,\text{GeV}$, except for those with small-x and different values of g_2, which are truncated at $p_T = 300\,\text{GeV}$.

6.8.2 Comparison of Data and ResBos

Figure 6.58 compares the corrected ϕ_η^* distributions with the predictions of RESBOS, separately for the different rapidity bins of the dielectron and dimuon channels. Figure 6.59 shows the ratio of the data to RESBOS. The yellow band indicates the uncertainty on the RESBOS prediction estimated by adding the PDF and scale uncertainties in quadrature. The renormalisation and factorisation scales are varied up and down by a factor of two. The blue line shows the RESBOS prediction, having chosen the value of g_2 that best describes the data. The black line shows the RESBOS prediction with the small-x broadening effect.

The general shape of the ϕ_η^* distribution is broadly described by RESBOS. However, the huge increase in precision compared to previous analyses exposes a failure of RESBOS to describe the data in detail. We make the following specific observations:

- Allowing the g_2 parameter to float has very little effect on the level of agreement.
- The prediction that includes small-x broadening is in even poorer agreement with the data and is excluded by the dielectron data for $|y| > 2$.

6.8.3 Fitting for g_2

Figures 6.60 and 6.61 shows, for the dielectron and dimuon channels respectively, fits for g_2, based on comparison of the unfolded distributions, with predictions from RESBOS with different g_2 values. The fits are restricted to $\pm 6\sigma$ around the minimum, in order to avoid regions with non-quadratic behaviour. The lower rows restrict the fits to the first 10 ϕ_η^* bins. It can be seen that the χ^2 per degree of freedom is typically large (e.g., 78/16 for the first $|y|$ bin in the dielectron channel).

In both the dielectron and dimuon data, the fitted values of g_2 show a monotonic *decrease* with increasing $|y|$. That is, the width of the ϕ_η^* distribution becomes narrower with increasing $|y|$ faster in the data than is predicted by default RESBOS. This is exactly the opposite of the behaviour expected from the small-x broadening hypothesis.

6.9 Cross Checks

6.9.1 Dielectron Versus Dimuon Comparison

Figure 6.59 compares the ratios of the unfolded dielectron and dimuon data to the appropriate RESBOS predictions. It can be seen that when compared in this way the dielectron and dimuon data are consistent with one another. Given that most experimental systematics are uncorrelated between the two channels this agreement provides a powerful cross check of the experimental method, corrections and systematic uncertainties presented in previous sections. The deficiencies of RESBOS in providing a detailed description of the shape of the ϕ_η^* distribution are confirmed in Fig. 6.59. It should be noted that since the dielectron and dimuon data are corrected to a slightly different physics level, Fig. 6.59 represents the most appropriate way to make a consistency check between the dielectron and dimuon data.

6.9.2 ϕ-Gap Checks

Figure 6.62 shows the ϕ_η^* distributions for both dielectron and dimuon channels, separately for events in which exactly 0, 1 and 2 leptons are in the CC-fiducial or EC regions of the detector. Figure 6.63 shows the ϕ_η^* distributions for both dielectron and dimuon channels, separately for events in which exactly 0, 1 and 2 leptons are in the fiducial muon acceptance (away from the octant boundaries). Figure 6.64 compares the relative fractions of the different categories in data and MC.

As a cross check of our modelling of the electron and muon ϕ-gaps, we can fit for the g_2 parameter at the *uncorrected* level. The detector level MC is re-weighted in p_T, and y, to match the predictions of RESBOS with different g_2 values, and a minimum χ^2 fit determines the value of g_2 which best describes the data. Note that since we have already applied an additional ϕ_η^* dependent K-factor to our MC events, the absolute value of g_2 cannot be compared with the fits in Sect. 6.8. The value of g_2 determined with this additional K-factor is referred to as g_2^{KF}. Table 6.1 compares the g_2^{KF} fits for different ϕ-gap and rapidity requirements, separately for the electron and muon channels. Since there are no significant discrepancies, electron and muon ϕ-gap efficiencies is considered to be adequately modelled.

For comparison, we also fit for g_2, without applying the additional ϕ_η^* dependent K-factor, as shown in the last two columns of Table 6.1. In this case, *we cannot expect stable fits*; the different ϕ-gap requirements sample different regions (see Figs. 6.62 and 6.63) of the ϕ_η^* distribution which is poorly modelled by RESBOS.

6.9.3 Unfolding Closure Test Using g_2

A powerful closure test of the unfolding method is to compare the values of g_2 obtained by:

1. Comparing the unfolded ϕ_η^* distributions directly with predictions from RESBOS/PHOTOS as in Sect. 6.8.
2. Comparing the unfolded ϕ_η^* distributions directly with predictions from PYTHIA re-weighted to RESBOS in p_T and y.

3. Comparing the un-corrected ϕ_η^* distributions with the fully simulated MC which has been re-weighted to RESBOS in p_T and y.

The fits are presented in Figs. 6.65 and 6.66 for the dielectron and dimuon channels respectively. It can be seen that the fits performed using PYTHIA before and after unfolding are consistent. The fits performed using RESBOS and PYTHIA are not necessarily expected to be consistent, since the two generators may predict a different translation between p_T and ϕ_η^*.

6.9.4 Data Subset Checks Using g_2

Table 6.2 shows the values of g_2 determined at detector level for various subsets of the data. The following subsets are considered: passing and failing the standard data quality; 0, 1 and 2 leptons with hits in the SMT; three different ranges of instantaneous luminosity; two different ranges of primary vertex locations along the beam direction; the four combinations of magnet polarity. Where appropriate, the same "special" requirements are made for data *and* MC. The exceptions are for the good and bad data quality, and the different magnet polarities, for which the special requirements are inapplicable in the MC.

References

1. C. Balazs, C.-P. Yuan, Phys. Rev. D **56**, 5558–5583 (1997)
2. E. Barberio, Z. Was, Comput. Phys. Commun. **79**, 291 (1994)
3. J.M. Butterworth et al., The Tools and Monte Carlo Working Group, Summary Report from the Les Houches 2009 Workshop on TeV Colliders, in *Les Houches 2009 Tools and Monte Carlo working group*, 2010. arXiv:1003.1643v1 [hep-ph]
4. T. Sjostrand, Comput. Phys. Commun. **135**, 238 (2001)
5. M.L. Mangano, JHEP **07**, 001 (2003)
6. M. Cooke, A. Croc, F. Déliot, H. Greenlee, A.S. Santos, V. Sharyy, B. Tuchming, M. Vesterinen, T. R.Wyatt. D0 Note 6031, D0, 2010
7. M. Vesterinen, T. R. Wyatt. D0 Note 6088, D0, 2010
8. V.M. Abazov et al., Phys. Rev. Lett. **106**, 122001 (2011)
9. P.M. Nadolsky et al., Implications of CTEQ global analysis for collider observables. Phys. Rev. D **78**, 013004 (2008)
10. F. Landry, R. Brock, P.N. Nadolsky, C.-P. Yuan, Fermilab tevatron run-1 z boson data and the collins-soper-sterman resummation formalism. Phys. Rev. D **67**, 073016 (2003)
11. S. Berge, P. Nadolsky, F. Olness, C.-P. Yuan, Phys. Rev. D **72**, 033015 (2005)

Chapter 7
Measurement of the ZZ and WZ Production Cross Sections

7.1 Introduction

We now move from the huge $Z/\gamma^* \to \ell^+\ell^-$ event samples, to the relatively low production cross section diboson processes. The processes, $ZZ/\gamma^* \to \nu\bar{\nu}\ell^+\ell^-$ and $WZ/\gamma^* \to \ell\nu\ell^+\ell^-$, are studied with the aim of making the most precise measurement of their cross sections.

In order to eliminate subjective bias in our analysis method, we performed a *blind* analysis. Selection requirements and analysis techniques were optimised based on MC expectation in the signal regions. Real data were however used in the verification of background modelling within signal-free control regions. Only once the event selection and background modelling were finalised did we look at the signal candidate events in real data.

7.2 Dataset and MC Samples

7.2.1 Dataset

The full Run II dataset is used, up to run number 270116 (18th March 2011). Unlike the ϕ_η^* analysis of Chap. 6, this analysis rejects periods of data taking which are marked as bad by the D0 data quality group. Whilst the ϕ_η^* analysis is rather insensitive to the effect of typical detector operation problems, the event signatures for $ZZ/\gamma^* \to \nu\bar{\nu}\ell^+\ell^-$ and $WZ/\gamma^* \to \ell\nu\ell^+\ell^-$ involve significant missing transverse momentum, which can be falsely generated by, e.g., noise in the calorimeter.

7.2.2 MC Samples

Table 7.1 lists the processes simulated in this analysis, either using PYTHIA [1] or ALPGEN [2] with showering and hadronisation provided by PYTHIA. The generator

Table 7.1 Table of processes simulated using MC

Process	Generator	$\sigma \times B$ (pb)	Generated events
$Z/\gamma^* \to e^+e^-$ ($15 < M_{Z/\gamma^*} < 60$)	PYTHIA	3.65e+02	1.84e+06
$Z/\gamma^* \to e^+e^-$ ($60 < M_{Z/\gamma^*} < 130$)	PYTHIA	1.76e+02	8.65e+07
$Z/\gamma^* \to e^+e^-$ ($130 < M_{Z/\gamma^*} < 250$)	PYTHIA	1.38e+00	2.73e+06
$Z/\gamma^* \to \mu^+\mu^-$ ($15 < M_{Z/\gamma^*} < 60$)	PYTHIA	3.65e+02	9.05e+06
$Z/\gamma^* \to \mu^+\mu^-$ ($60 < M_{Z/\gamma^*} < 130$)	PYTHIA	1.76e+02	3.82e+07
$Z/\gamma^* \to \mu^+\mu^-$ ($130 < M_{Z/\gamma^*} < 250$)	PYTHIA	1.38e+00	2.92e+05
$Z/\gamma^* \to \tau^+\tau^-$ ($60 < M_{Z/\gamma^*} < 130$)	PYTHIA	1.82e+02	3.06e+06
$Z/\gamma^* \to \tau^+\tau^-$ ($130 < M_{Z/\gamma^*} < 250$)	PYTHIA	1.35e+00	8.02e+05
$WW \to \ell^+\nu\ell^-\bar{\nu}$	PYTHIA	8.38e−01	8.31e+05
$WZ/\gamma^* \to \ell\nu\ell^+\ell^-$	PYTHIA	7.84e−02	3.63e+05
$Z/\gamma^*Z/\gamma^* \to \ell^+\ell^-\ell^+\ell^-$	PYTHIA	6.75e−02	4.84e+05
$ZZ/\gamma^* \to \nu\bar{\nu}\ell^+\ell^-$	PYTHIA	6.75e−02	4.84e+05
$W\gamma \to e\nu\gamma$	PYTHIA	2.68e+00	9.52e+05
$Z\gamma \to ee\gamma$	PYTHIA	1.06e+00	8.75e+05
$W \to e\nu$	PYTHIA	1.95e+03	1.89e+07
$W \to \mu\nu$	PYTHIA	1.95e+03	1.89e+07
$t\bar{t} \to ll\nu\nu bb$ (+0lp)	ALPGEN+PYTHIA	3.56e−01	7.50e+05
$t\bar{t} \to ll\nu\nu bb$ (+1lp)	ALPGEN+PYTHIA	1.43e−01	4.52e+05
$t\bar{t} \to ll\nu\nu bb$ (+2lp)	ALPGEN+PYTHIA	7.14e−02	2.82e+05

The $\sigma \times B$, and number of generated events are listed

Fig. 7.1 *Colour codes* used throughout this chapter

name, $\sigma \times B$ from the generator, and number of generated events are provided. Figure 7.1 shows the colour scheme used to distinguish the different processes in this chapter.

7.3 Dilepton Preselection

The first stage of the analysis selects an inclusive sample of dilepton events, which is dominated by $Z/\gamma^* \to \ell^+\ell^-$. The lepton quality requirements are tighter for the ZZ/γ^* dilepton preselection than in the WZ/γ^* dilepton preselection, due to the larger instrumental background from W+jet events in the ZZ/γ^* channel. Otherwise, the two analyses share the same dilepton preselection requirements.

7.3 Dilepton Preselection

In addition to the ee and $\mu\mu$ decay channels, we make use of the $e\mu$ channel in the ZZ/γ^* analysis. The final $ZZ/\gamma^* \to \nu\bar{\nu}\ell^+\ell^-$ candidate sample will be dominated by $WW \to \ell^+\nu\ell^-\bar{\nu}$ which has a significant branching fraction into the $e\mu$ channel. The $e\mu$ channel therefore serves as a signal-free control channel that is useful in verifying the accuracy with which this background is modelled.

7.3.1 Trigger Requirements

The previous analyses of these channels [3, 4] required events to have fired single-lepton triggers. In order to maximise our overall signal efficiency, we make no specific trigger requirements in offline analysis. Most of the analysed events are still recorded based on single-lepton or dilepton triggers. However, additional efficiency is gained through, e.g., lepton+jet triggers. For the ZZ/γ^* analysis, the trigger efficiency is mostly cancelled in the ratio of signal to inclusive Z/γ^*. For the WZ/γ^* analysis, there is a bias of roughly 5 % in the $Z/\gamma^* \to \mu^+\mu^-$ channel due to the additional lepton in the signal sample. This bias is corrected for and introduces no significant uncertainty as discussed in Sect. 7.11. There is no significant bias in the $Z/\gamma^* \to e^+e^-$ channel, since the single-electron and dielectron triggers are close to 100 % efficient for the dilepton selection.

7.3.2 Lepton Quality Definitions

Loose, medium and tight qualities are defined for electrons and muons. The specific requirements are listed in Table 7.2. The loose definitions are only used for background estimation; medium definitions are used for the Z/γ^* daughters in the WZ/γ^* analysis; tight definitions are used for the Z/γ^* daughters in the ZZ/γ^* analysis and the W daughter in the WZ/γ^* analysis. Electrons are treated differently depending on whether they are reconstructed in the CC, EC or IC regions of the detector.

7.3.3 $ZZ/\gamma^* \to \nu\bar{\nu}\ell^+\ell^-$ Dilepton Preselection Requirements

The basic selection is an opposite charge pair of high p_T leptons. Same charge dilepton events are considered as a control sample. A primary vertex is defined by the two candidate lepton tracks. In events where exactly one of the tracks has SMT hits, the longitudinal coordinate of this vertex, z_{pv}, is defined to be equal to the z of that track (measured from the detector centre). Otherwise, z_{pv} is defined as the mean of the z coordinates of the two tracks. We require that $|z_{\text{pv}}| < 80\,\text{cm}$, and that the z coordinates of the lepton tracks are within 3 cm of each other. The dilepton invariant mass, $M_{\ell\ell}$, must be between 60 and 120 GeV. The regions, $40 < M_{\ell\ell} < 60\,\text{GeV}$ and $M_{\ell\ell} > 120\,\text{GeV}$ are considered as control samples. The following requirements are specific to the different decay channels:

Table 7.2 Electron and muon quality cut definitions

	Loose	Medium	Tight
CC electrons			
$|\eta_{det}|$	<1.1	<1.1	<1.1
BDT output	>−2	>−0.9	>−0.8
EC electrons			
$|\eta_{det}|$	1.5–3	1.5–3	1.5–3
BDT output	>−2	>−0.98	>−0.7
IC electrons			
$|\eta_{det}|$	1.1–1.5	1.1–1.5	1.1–1.5
χ^2_{trk}	<9.5	<9.5	<9.5
n_{SMT}	≥4	≥4	≥4
n_{CFT}	≥11	≥11	≥11
$\mathcal{I}^{hc4}_{trk}/p_T$	–	–	<0.2
$NN_\tau(e)$	>0.0	>0.9	>0.95
Muons			
μ-quality	Loose	Loose	Loose
$|\eta_{det}|$	<2.1	<2.1	<2.1
χ^2_{trk}	<4.0	<4.0	<4.0
$|r_{dca}|(n_{SMT}=0)$	<0.2 cm	<0.2 cm	<0.2 cm
$|r_{dca}|(n_{SMT}>0)$	<0.015 cm	<0.015 cm	<0.015 cm
$\mathcal{I}_{trk}/p_T$	<0.4	<0.25	<0.1
$\mathcal{I}_{cal}/p_T$	<0.4	<0.25	<0.1

The usage of these definitions is given in the text

- ***ee* channel** One tight (CC/EC) electron with $p_T > 20$ GeV, and one other opposite charge tight (CC/EC) electron or tight type2 IC-electron with $p_T > 15$ GeV. Type-2 IC electrons have a reasonable energy resolution, whereas type-1 IC electrons rely on the central track which has a relatively poor p_T resolution. Rejecting mis-measured Drell-Yan events with type-1 IC electrons would be a significant challenge, for a very small gain in acceptance.
- ***μμ* channel** One tight muon with $p_T > 15$ GeV, and another tight muon of opposite charge and with $p_T > 10$ GeV. In order to reject cosmic ray muon events, we require that $|\eta_1 + \eta_2| > 0.05$.
- ***eμ* channel** A tight muon and a tight (CC/EC) electron[1] with opposite charge. We attempt to mimic the asymmetric lepton p_T requirements of the *ee* and *μμ* channels. Both leptons must satisfy the softer p_T cut (10 GeV for muons and 15 GeV for electrons), and at least one must satisfy the harder p_T cut (15 GeV for muons and 20 GeV for electrons).

---

[1] The data analysis starts from a skim of the dataset that is commonly used for analysing $e\mu$ final states. Unfortunately, this skim requires a CC or EC electron. These requirements were defined before the development of electron identification in the IC region.

7.3 Dilepton Preselection

7.3.4 $WZ/\gamma^* \to \ell\nu\ell^+\ell^-$ Dilepton Preselection Requirements

As for the ZZ/γ^* preselection, except that the tight lepton quality requirements are relaxed to medium, and type-2 IC electrons are allowed in the ee channel in addition to type-1 IC electrons.

7.4 Additional Objects

7.4.1 Jets

Section 7.7 introduces special missing transverse momentum estimators to separate the $ZZ/\gamma^* \to \nu\bar{\nu}\ell^+\ell^-$ signal from the Drell-Yan background. The construction of these variables makes extensive use of information on reconstructed jets; both in the calorimeter and the tracker. In addition, the ZZ/γ^* selection cuts (see Sect. 7.10) include a veto on the presence of more than two calorimeter jets. The requirements on reconstructed calorimeter jets and track jets are as follows:

- **Calorimeter jets** Jets are reconstructed with the D0 midpoint cone algorithm with a cone size of $\Delta\mathcal{R} < 0.5$ as described in Chap. 3. We require jets to have $p_T > 15\,\text{GeV}$, and to be separated by $\Delta\mathcal{R} > 0.3$ from the leptons that are assigned to the $Z/\gamma^* \to \ell^+\ell^-$ decay.
- **Track jets** We select a collection of tracks that satisfy $p_T > 1\,\text{GeV}$ and are separated by $\Delta\mathcal{R} > 0.3$ from the leptons. Tracks must have a z coordinate that is consistent with the primary vertex within 1 cm, and further satisfy: $\chi^2/\text{NDF} < 4$ and $r_{\text{dca}} < 0.2\,\text{cm}$. Track jets are reconstructed using a simple cone algorithm. The highest p_T track in the collection is taken as a seed, and is combined with all other tracks in the collection within $\Delta\mathcal{R} < 0.5$. Tracks are removed from the collection once they have been combined into a jet. This process continues until no tracks remain in the collection. Track jets are required to contain at least two tracks.

7.4.2 Missing Transverse Energy

The missing transverse energy, $\slashed{E}_T$, is calculated as described in Chap. 3. Coarse hadronic cells are excluded to minimise the effect of noise. Corrections are applied for leptons that satisfy the loose requirements. The "raw" calorimeter E_T for each lepton is added vectorially to the $\slashed{E}_T$. For muons, this is based on the estimated energy loss. For electrons, this is based on the measured energy before energy loss corrections. The p_T used in kinematic analysis of the leptons (e.g., central track p_T for muons) is subtracted vectorially from the $\slashed{E}_T$.

7.4.3 Additional Leptons

Two potential sources of background in the ZZ/γ^* analysis are the processes, $WZ/\gamma^* \to \ell\nu\ell^+\ell^-$ and $Z/\gamma^*Z/\gamma^* \to \ell^+\ell^-\ell^+\ell^-$. These backgrounds are suppressed by vetoing on the presence of additional leptons, other than the pair assigned to the $Z/\gamma^* \to \ell^+\ell^-$ decay. Similarly, $Z/\gamma^*Z/\gamma^* \to \ell^+\ell^-\ell^+\ell^-$ is a background in the WZ/γ^* analysis, which we suppress by vetoing on the presence of a fourth lepton. The full selection requirements for the ZZ/γ^* and WZ/γ^* analyses are described in Sects. 7.10 and 7.11 respectively.

The previous D0 analysis [3] of the $ZZ/\gamma^* \to \nu\bar{\nu}\ell^+\ell^-$ process allowed these additional "veto" leptons to be of substantially loose quality than those of the $Z/\gamma^* \to \ell^+\ell^-$ decay. Reconstructed EM clusters, muons, taus, and isolated tracks are all considered. The "veto" lepton objects must be separated from the leptons assigned to $Z/\gamma^* \to \ell^+\ell^-$ decay by $\Delta\mathcal{R} > 0.2$. Additional quality requirements are defined as follows:

- **EM clusters** EM clusters must satisfy $p_T > 5\,\text{GeV}$, and either (a) be matched to a central track with $p_T > 8\,\text{GeV}$ and $|r_{\text{dca}}| < 0.1\,\text{cm}$, or (b) satisfy basic shower shape requirements ($\chi^{2(7)}_{\text{EM}} < 12$ for CC clusters and $\chi^{2(8)}_{\text{EM}} < 20$ for EC clusters).
- **Muons** Loose quality local muons must be matched to a central track. Medium quality local muons are considered even if they do not have a central track match. There are no specific p_T or isolation requirements. Poorly isolated muons may indicate a semi-leptonic decay of a b-flavour hadron. Vetoing on such muons is therefore likely to suppress the background from $t\bar{t}$ production.
- **Type-3 taus** Taus must satisfy $p_T > 5\,\text{GeV}$, and a have neural network output of at least 0.3. They must also contain a central track whose longitudinal coordinate is consistent with the primary vertex within 1 cm. Single-prong hadronic tau decays (typically reconstructed as type-1 or type-2 taus) are expected to be picked up as isolated tracks (next bullet).
- **Isolated tracks** Isolated tracks must satisfy $p_T > 5\,\text{GeV}$. The isolation requirement is that the calorimeter E_T within an isolation cone of size $\Delta\mathcal{R} < 0.7$ must be no larger than the track p_T.

7.5 Corrections to the Simulation

A number of corrections are applied to the simulated MC event samples in order to improve the accuracy with which they are able to describe the data.

- **Instantaneous luminosity profile** Events with $25 < \mathcal{L}_{\text{inst}} < 100 \times 10^{30}\,\text{cm}^{-2}\,\text{s}^{-1}$ are scaled down by a factor of two.
- **$Z/\gamma^* p_T$ reweighting** The PYTHIA $Z/\gamma^* \to \ell^+\ell^-$ events are reweighted in two dimensions (dilepton p_T and rapidity) to the predictions of RESBOS [5],[2] as in

---

[2] We use the CP version of the code and grid files.

7.5 Corrections to the Simulation

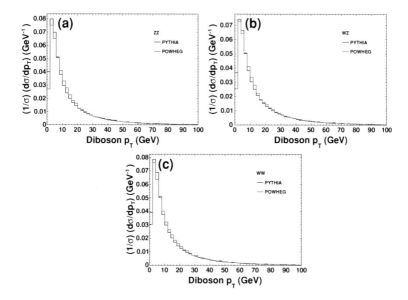

Fig. 7.2 Comparison of PYTHIA and POWHEG predictions for the diboson p_T distributions

the ϕ_η^* analysis of Chap. 6. The g_2 parameter is set to the values measured in that analysis, separately in the three rapidity bins.

- **Diboson p_T reweighting** The PYTHIA diboson MC events are reweighted in diboson p_T to the predictions of the NLO MC program POWHEG [6, 7]. Figure 7.2 compares the shapes of the p_T distributions predicted by PYTHIA and POWHEG.
- **Calorimeter jets** Corrections are applied for differences between simulation and data in reconstruction efficiency and energy resolution [8]. An additional energy offset of -1 GeV is applied to jets in the ICR region in order to improve the agreement in the shape of the jet η distributions.
- **Electron energies** CC and EC electron energies receive the treatment derived in Chap. 4. For type-1 IC electrons, we use the central track p_T. The track p_T is smeared with a similar form to that used for muons [9], but with an additional energy loss correction [10]. For type-2 IC electrons, we use the calorimeter energy which provides significantly better resolution than the tracker. A simple energy smearing is applied to the MC [11].
- **Muon p_T smearing** A commonly used smearing is applied to the p_T of muons [9].
- **Electron and muon reconstruction and selection efficiencies** The efficiencies for all electron and muon reconstruction and selection requirements are measured in data and MC using similar methods to those used in the ϕ_η^* analysis. Figures 7.3 and 7.4 show the data-vs-MC efficiency ratios for requirements relevant to electrons and muons respectively.

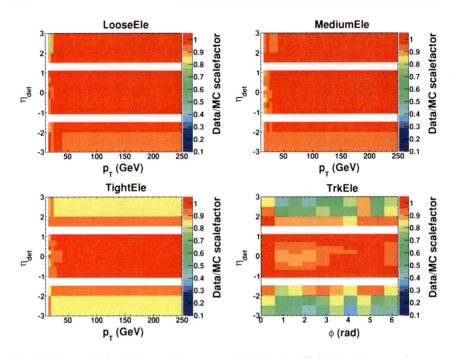

Fig. 7.3 Data-vs-MC scale-factors for reconstruction/selection efficiencies relevant to electrons. Clockwise from *top left*: loose electron reconstruction efficiency; medium electron efficiency with respect to loose elections; tight electron efficiency with respect to loose elections; electron track reconstruction efficiency

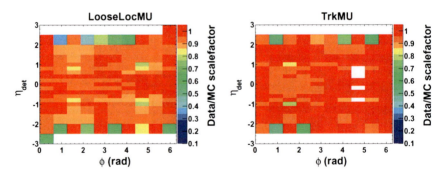

Fig. 7.4 Data-vs-MC scale-factors for reconstruction/selection efficiencies relevant to muon. *Left* loose local muon reconstruction efficiency. *Right* muon track reconstruction efficiency

- **Electron track p_T smearing** The same prescription is used as derived for the ϕ_η^* analysis of Chap. 6. The central track p_T is used in an estimate of electron E_T mis-measurement for rejection of Drell-Yan background in the ZZ/γ^* analysis (see Sect. 7.7).

7.5 Corrections to the Simulation

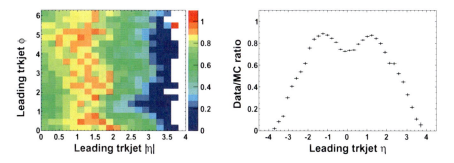

Fig. 7.5 Data-vs-MC ratio as a function of leading trackjet (*left*) $|\eta|$ and ϕ, and (*right*) η, at the dilepton preselection stage

- **Trackjet reconstruction efficiency** Figure 7.5 shows the data-vs-MC ratio, R_{trkjet}, as a function of the leading trackjet $|\eta|$ and ϕ. Trackjets are randomly removed with a probability of $1 - R_{\text{trkjet}}$. Figure 7.5 also shows R_{trkjet} as a function of η.
- **Trigger efficiency correction** A trigger efficiency correction is necessary for the $Z/\gamma^* \to \mu^+\mu^-$ channels, but only affects the WZ/γ^* analysis. This correction accounts for the increased trigger efficiency in dimuon events with an additional high p_T electron or muon. A more detailed explanation of this correction is provided in Sect. 7.11.

7.6 Comparison of Data and Simulation After Dilepton Selection

It is important to verify that the simulation accurately describes the basic kinematic distributions after the dilepton preselection. Figures 7.6, 7.7, 7.8, 7.9 and 7.10 compare simulation and data after the ZZ/γ^* dilepton preselection, i.e., requiring two tight quality leptons. The dilepton invariant mass distribution is shown without imposing the requirement of $60 < M_{\ell\ell} < 120$ GeV. The total MC prediction has been normalised to the data in the region $60 < M_{\ell\ell} < 120$ GeV, separately for the ee, $\mu\mu$ and $e\mu$ channels. This implies that we only rely on the MC event generators for the *ratios* of cross sections for the different processes considered.

In this section, the lower half of each plot shows the ratio of data to the MC prediction, having scaled both to the same area, thus *only taking into account the shape information*. The yellow band represents the total systematic uncertainty on the MC prediction. The red and blue lines represent the two dominant systematic uncertainties for the plotted distribution. The ranking of systematics is as follows: the modulus of the deviation of the ratio from unity is calculated for each bin, and then a sum is formed over the bins. The systematic uncertainties are discussed in more detail in Sect. 7.14. In general, the level of agreement between simulation and data is considered to be sufficient for the purposes of this analysis. As we shall see in

150   7 Measurement of the ZZ and WZ Production Cross Sections

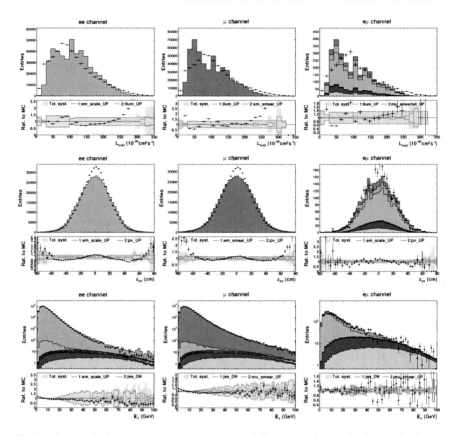

Fig. 7.6 Data-vs-MC comparison of the $\mathcal{L}_{\text{inst}}$ z_{pv} and $\slashed{E}_T$ distributions at the (ZZ/γ^*) dilepton stage

Sect. 7.14, systematic uncertainties are far outweighed by the dominating statistical uncertainties on the signal cross section measurements.

Certain distributions are rather poorly modelled, even within the estimated systematic uncertainties. For example the jet p_T distributions are in poor agreement. Fortunately, the analysis is designed to have minimal sensitivity to the precise modelling of the jets. The most important use of the reconstructed jets is in the construction of the missing p_T estimators (see Sect. 7.7) that are needed to select ZZ/γ^* candidate events. This only requires us to reconstruct the total p_T of the hadronic recoil and is not particularly sensitive to the properties of the individual jets. The dielectron invariant mass distribution is in rather poor agreement, with an excess of around 10 % in the region, $M_{\ell\ell} < 75$ GeV. Since this is not covered by any reasonable variation in the energy resolution, the most likely explanation is a missing multijet background component. This background cannot be more than 1 % or so, in the Z peak region, which would have a negligible effect on the cross section measurement.

7.7 Missing Transverse Momentum Estimators

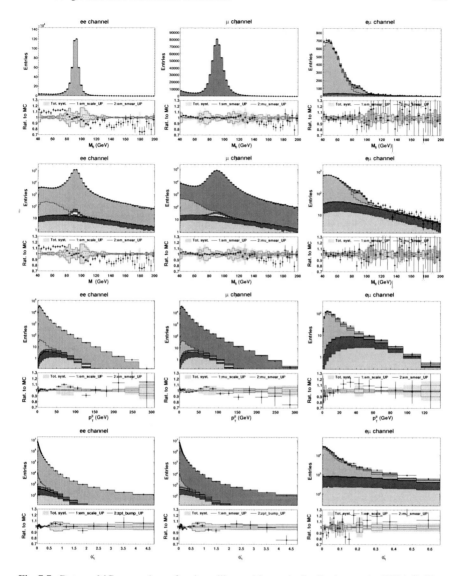

Fig. 7.7 Data-vs-MC comparison of various dilepton kinematic distributions at the (ZZ/γ^*) dilepton stage

7.7 Missing Transverse Momentum Estimators

The basic signature of $ZZ/\gamma^* \to \nu\bar{\nu}\ell^+\ell^-$ is a pair of charged leptons with an invariant mass around m_Z, produced in association with significant missing transverse momentum, $\not{p}_T$ due to the neutrinos from the $Z \to \nu\bar{\nu}$ decay. A substantial background is possible from inclusive $Z/\gamma^* \to \ell^+\ell^-$ production in which the

152   7 Measurement of the ZZ and WZ Production Cross Sections

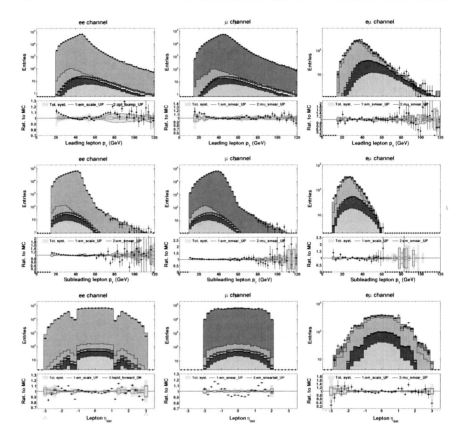

Fig. 7.8 The *top* and *middle rows* show the leading and sub-leading lepton p_T distributions respectively, at the (ZZ/γ^*) dilepton stage. The *bottom row* shows the lepton η_{det} distribution (with two entries per event)

leptons and/or any hadronic recoil is mis-reconstructed. Very stringent requirements are required since (i) the production cross section for Z/γ^* exceeds that of the signal by four orders of magnitude and (ii) the rates of gross mis-reconstruction are simulated with limited accuracy.

Rather than make an estimate of the genuine $\not{p}_T$ in the event, we follow the approach of the previous D0 analysis of this process [3] and construct variables that represent the minimum $\not{p}_T$ that is feasible given the measurement uncertainties on the leptons and the hadronic recoil.

7.7.1 Construction of the Variables, $\not{p}'_T \not{q}'_T \not{q}'_L$

Firstly, the dilepton p_T is decomposed into a_T and a_L components with respect to the thrust axis as described in Chap. 5. The dilepton p_T is denoted p_T^{dilep}, and

7.7 Missing Transverse Momentum Estimators

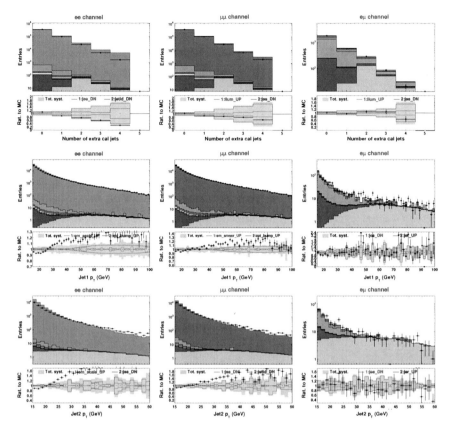

Fig. 7.9 The *top row* shows the jet multiplicity distribution at the ZZ/γ^* dilepton stage. The *middle* and *bottom rows* show the leading and sub-leading jet p_T distributions

the components as a_T^{dilep} and a_L^{dilep}. As discussed at length in Chap. 5, in the region $\Delta\phi > \pi/2$, the a_T component is relatively insensitive to lepton p_T mis-measurement. For $\Delta\phi < \pi/2$, this decomposition no longer make sense, and a_T^{dilep} and a_L^{dilep} are set equal to p_T^{dilep}. The distribution of these variables is shown in Fig. 7.11.

We attempt to reconcile the apparent magnitude of each of these components with (i) lepton p_T mis-measurement, (ii) a reconstructed hadronic recoil in the calorimeter, and (iii) any remaining recoil activity in the tracking detector.

Lepton p_T Mis-Measurement

A correction for possible lepton p_T mis-measurement is determined by floating the individual lepton p_Ts within one standard deviation of their estimated uncertainties in order to separately minimise $a_T^{\text{dilep}}, a_L^{\text{dilep}}$ and p_T^{dilep}. The transfer functions used to estimate the p_T uncertainties are determined in Appendix A. The a_T^{dilep} is minimised

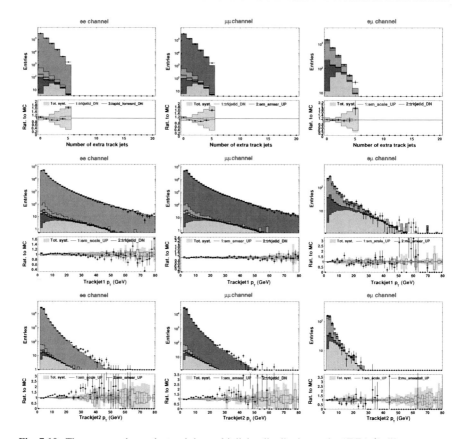

Fig. 7.10 The *top row* shows the track jet multiplicity distribution at the (ZZ/γ^*) dilepton stage. The *middle* and *bottom rows* show the leading and sub-leading track jet p_T distributions

by floating the p_T of both leptons *down* by one standard deviation. The a_L^{dilep} floats the p_T of the leading lepton *down*, but the p_T of the subleading lepton *up*. The p_T^{dilep} is minimised by choosing the variation (down-down or up-down) that gives the largest reduction. Electrons that are close to module boundaries in the CC or are in the IC have relatively poor energy resolution and are given special treatment. If the p_T measured by the tracker is *larger* than the E_T measured by the calorimeter then this variation is considered as another option in the minimisation. Correction terms are defined, e.g., for a_T^δ:

$$a_T^\delta = \text{Min}(0, a_T^{\text{reduced}} - a_T^{\text{dilep}}),$$

where a_T^{reduced} is the result of minimising a_T^{dilep}. Figure 7.12 shows the p_T^δ, a_T^δ and a_L^δ distributions. In the few cases where the minimisation "overshoots" (indicated by positive values), we set, e.g., $a_L^\delta = -a_L^{\text{dilep}}$.

7.7 Missing Transverse Momentum Estimators

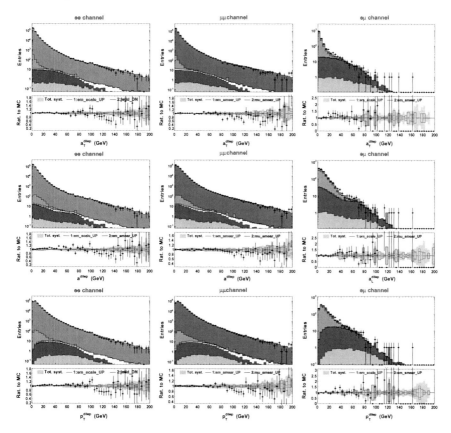

Fig. 7.11 Data-vs-MC comparison of the a_T^{dilep}, a_L^{dilep}, and p_T^{dilep} distributions after dilepton preselection

7.7.2 Calorimeter Recoil

Two estimates of the calorimeter recoil are made: one from the reconstructed jets, and another from the reconstructed $\not{E}_T$. The p_T, a_T and a_L components are calculated for each jet in the event, e.g., $a_T^{\text{jet}(i)} = |\vec{p}_T^{\text{jet}(i)} \times \hat{t}|$, where $\vec{p}_T^{\text{jet}(i)}$ is the p_T vector of the ith jet. The most negative combination of the jets is constructed; e.g., for the a_T component:

$$a_T^{\text{jets}} = \text{Min}(0, a_T^{\text{jet}(1)}) + \text{Min}(0, a_T^{\text{jet}(2)}) + \cdots + \text{Min}(0, a_T^{\text{jet}(n)}).$$

This approach ensures that jets which are not genuinely associated with the recoil system (e.g. from additional $p\bar{p}$ collisions or calorimeter noise) are not allowed to generate a fake imbalance in an otherwise well reconstructed event.

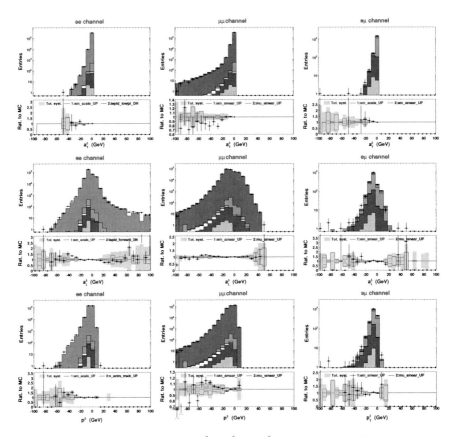

Fig. 7.12 Data-vs-MC comparison of the a_T^δ !!, a_L^δ, and p_T^δ distributions after dilepton preselection

The $\slashed{E}_T$ estimate subtracts any contribution from the two leptons and then tests how the remaining $\slashed{E}_T$ balances with the dilepton system. Of the jet and $\slashed{E}_T$ corrections, we take the one with the most negative value in determining a combined calorimeter correction; e.g. for the a_T component:

$$a_T^{\text{recoil}} = \text{Min}(0, a_T^{\text{jets}}, a_T^{\text{met}}),$$

where a_T^{met} is the $\slashed{E}_T$ based correction.

Track Recoil

As a protection against possible failure to reconstruct a recoiling hadronic system in the calorimeter, we attempt to recover any remaining activity in the tracker. Up to four track jets (see Sect. 7.4) are considered. Corrections to the each of the (p_T, a_T and a_L) components are determined in the same way as for calorimeter jets; e.g., for the a_T component (Figs. 7.13, 7.14, 7.15):

7.7 Missing Transverse Momentum Estimators

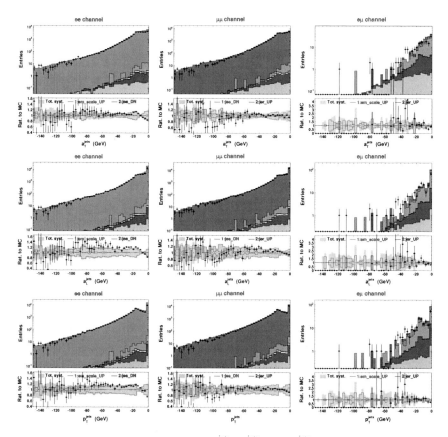

Fig. 7.13 Data-vs-MC comparison of the a_T^{jets}, a_L^{jets}, and p_T^{jets} distributions after dilepton preselection

$$a_T^{\text{trkjets}} = \text{Min}(0, a_T^{\text{trkjet}(1)}) + \text{Min}(0, a_T^{\text{trkjet}(2)}) + \cdots + \text{Min}(0, a_T^{\text{trkjet}(n)}).$$

Figure 7.16 shows the p_T^{trkjets}, a_T^{trkjets} and a_L^{trkjets} distributions.

Combination

The missing transverse momentum estimators, p'_T, q'_T, and q'_L are constructed as follows:

$$p'_T = p_T^{\text{dilep}} - 2\left[p_T^{\delta} - \text{Min}(0, p_T^{\text{recoil}}) - \text{Min}(0, p_T^{\text{trkjets}})\right],$$

$$q'_T = a_T^{\text{dilep}} - 2\left[a_T^{\delta} - \text{Min}(0, a_T^{\text{recoil}}) - \text{Min}(0, a_T^{\text{trkjets}})\right],$$

$$q'_L = a_L^{\text{dilep}} - 2\left[a_L^{\delta} - \text{Min}(0, a_L^{\text{recoil}}) - \text{Min}(0, a_L^{\text{trkjets}})\right].$$

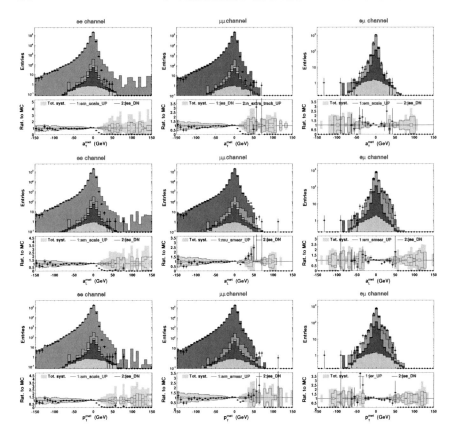

Fig. 7.14 Data-vs-MC comparison of the a_T^{met}, a_L^{met}, and p_T^{met} distributions after dilepton preselection

The factor of two adds further conservatism, and is found to be optimal based on MC simulation. Additional scalings of the different correction terms are unable to significantly improve the performance.

7.7.3 Weighted Combination of q'_T and q'_L

The previous D0 analysis of the $ZZ/\gamma^* \to \nu\bar{\nu}\ell^+\ell^-$ process used a weighted combination of q'_T and q'_L as the selection variable, denoted $\mathcal{E}'_T$. This is essentially a weighted quadrature sum, that gives more weight to the a_T component. In order to have a meaningful definition of this variable for negative values, we choose the following prescription:

7.7 Missing Transverse Momentum Estimators

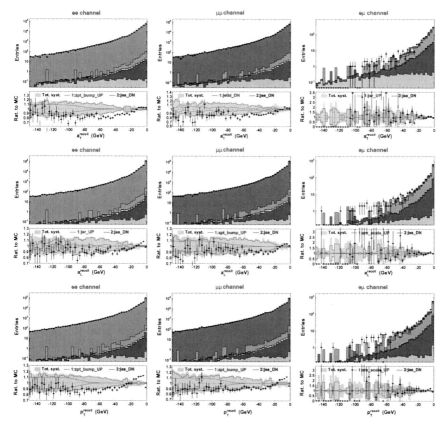

Fig. 7.15 Data-vs-MC comparison of the a_T^{recoil}, a_L^{recoil}, and p_T^{recoil} distributions after dilepton preselection

$$\not{E}'_T = \begin{cases} -\left[q'^2_L + 1.5 q'^2_T\right]^{\frac{1}{2}} & \text{if } q'_T < 0 \text{ and } q'_L < 0, \\ \left[q'^2_L + 1.5 q'^2_T\right]^{\frac{1}{2}} & \text{if } q'_T > 0 \text{ and } q'_L > 0, \\ \left[\text{Min}(0, q'_L)^2 + 1.5 \text{Min}(0, q'_T)^2\right]^{\frac{1}{2}} & \text{if } q'_T \times q'_L < 0. \end{cases}$$

7.7.4 Comparison of the Discriminating Variables

Figure 7.17 shows the p'_T, q'_T, q'_L and $\not{E}'_T$ distributions. In presenting the $e\mu$ data we stress that our aim is not to make a selection of $e\mu$ ($+\not{p}_T$) events with the highest possible performance in terms of efficiency/background discrimination. The $e\mu$ channel is dominated by sources of events with genuine missing p_T (WW and $t\bar{t}$) and has little background from Drell-Yan. These variables could therefore be regarded as

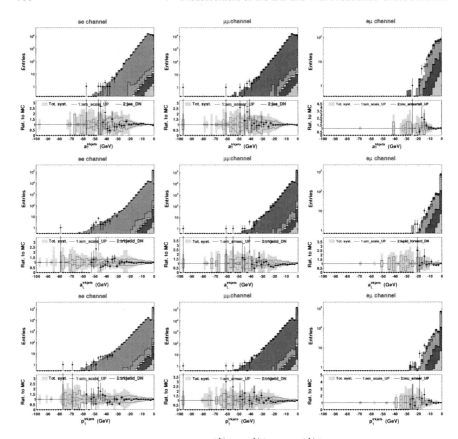

Fig. 7.16 Data-vs-MC comparison of the a_T^{trkjets}, a_L^{trkjets}, and p_T^{trkjets} distributions after dilepton preselection

"overkill" in this channel. Rather, our aim is to use the $e\mu$ channel to cross check the performance of our analysis in events with genuine $\not{p}_T$ and also test our predictions for non-Drell Yan backgrounds, such as W+jets. The performance of the different variables in separating $ZZ/\gamma^* \to \nu\bar{\nu}\ell^+\ell^-$ signal from mis-measured Drell-Yan will be evaluated in the next section.

7.7.5 Performance of the Variables

Figure 7.18 compares the background ($Z/\gamma^* \to \ell^+\ell^-$ in all lepton flavours) efficiency versus signal ($ZZ/\gamma^* \to \nu\bar{\nu}\ell^+\ell^-$) efficiency curves for the variables, $\not{p}_T'$, $\not{q}_T'$, $\not{q}_L'$, $\not{H}_T'$, and the standard $\not{H}_T$. As shall be demonstrated later, the most interesting region is background efficiencies below 10^{-5} or so. In the ee channel, the $\not{H}_T'$

7.7 Missing Transverse Momentum Estimators

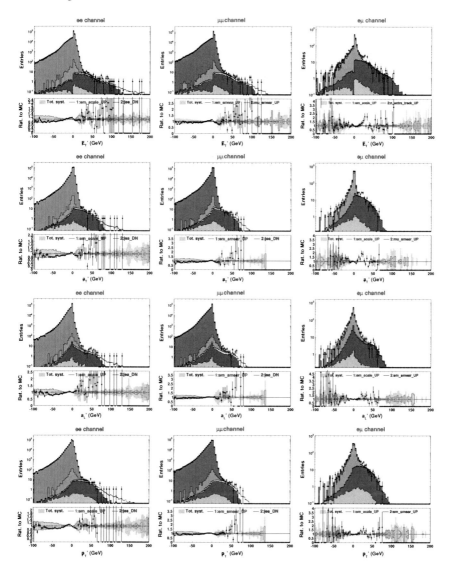

Fig. 7.17 Data-vs-MC comparison of the $\rlap{/}{E}'_T$, q'_T, q'_L, and $\rlap{/}{p}'_T$ distributions after dilepton preselection

and $\rlap{/}{p}'_T$ variables give the best performance, with little difference between the two. The standard $\rlap{/}{E}_T$ variable performs very poorly, with a maximum rejection power that is many orders of magnitude poorer than the other variables. In the $\mu\mu$ channel, the $\rlap{/}{E}'_T$, $\rlap{/}{p}'_T$, and q'_T variables give a similar signal efficiency at a background efficiency around 0.5×10^{-5} or so. However, at smaller background efficiencies, the q'_T variable significantly outperforms all other variables. The q'_L variable is less robust

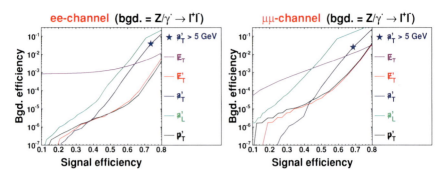

Fig. 7.18 Comparison of background efficiency versus signal efficiency *curves* for candidate variables that discriminate between Drell-Yan and ZZ/γ^* signal

against lepton p_T mis-measurement compared to $\not{q}'_T$. The difference in performance is noticeably larger in the $\mu\mu$ channel, due to the relatively poor muon p_T resolution.

A full description of the ZZ/γ^* selection cuts is given in Sect. 7.10. Here, we only discuss the selection of significant $\not{p}_T$. It is decided to apply require $\not{q}'_T > 5$ GeV by default, as indicated by the star shaped marker in Fig. 7.18. This "soft" $\not{q}'_T$ cut alone does not achieve the necessary rejection against $Z/\gamma^* \to \ell^+\ell^-$, but has the advantage of effectively eliminating the background from $Z/\gamma^* \to \tau^+\tau^-$ altogether.[3] Figure 7.19 shows the $Z/\gamma^* \to \tau^+\tau^-$ efficiency versus signal efficiency curves. The $\not{q}_T$ variable gives the highest performance for $Z/\gamma^* \to \tau^+\tau^-$ efficiencies below 10^{-2} in the *ee* channel and below 10^{-1} in the $\mu\mu$ channel. Whilst $Z/\gamma^* \to \tau^+\tau^-$ requires a much smaller rejection due to the small branching fraction into ee and $\mu\mu$ final states and the relatively low acceptance in lepton p_T, it typically has a genuine missing p_T, most of which is along the a_L direction.

Figure 7.20 shows the efficiency curves for the combination of this soft $\not{q}'_T$ cut and an additional cut on one of the candidate variables. This time, the background is $Z/\gamma^* \to \ell^+\ell^-$. The $\not{p}'_T$ and $\not{E}'_T$ variables give the best performance. With little else to chose between the two, the $\not{p}'_T$ variable is chosen for its simpler definition, and marginally better performance in the *ee* channel. A cut of $\not{p}'_T > 30$ GeV is indicated by the star shaped marker. This cut is optimised based on the expected signal cross section uncertainty including statistical and systematic sources as will be described in Sect. 7.7.

We further study the dependence of the performance on the qualities (in terms of resolution) of the leptons. Two categories are defined for each of the *ee* and $\mu\mu$ channels.

- **Category-1** Both leptons have *good* resolution: muons with SMT hits and with $|\eta_{\text{det}}| < 1.6$ and electrons within the fiducial CC or EC.

---

[3] This is also convenient since the available $Z/\gamma^* \to \tau^+\tau^-$ MC samples are relatively small (see Sect. 7.2) and could potentially introduce unsightly statistical fluctuations at the selected signal candidate stage of the analysis.

7.7 Missing Transverse Momentum Estimators

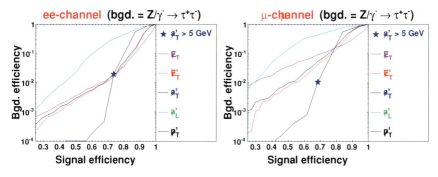

Fig. 7.19 Comparison of background efficiency versus signal efficiency *curves* for candidate variables that discriminate between Drell-Yan and ZZ/γ^* signal. The background process is $Z/\gamma^* \to \tau^+\tau^-$ only

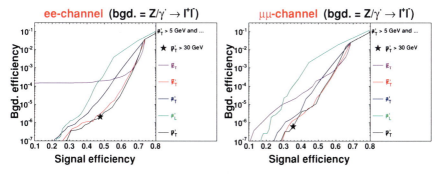

Fig. 7.20 Comparison of background efficiency versus signal efficiency curves for candidate variables that discriminate between Drell-Yan and ZZ/γ^* signal. The "soft" q'_T cut is included

- **Category-2** At least one lepton does not have *good* resolution.

Figure 7.21 is equivalent to Fig. 7.20 except splitting into the categories defined above. The $p'_T > 30\,\text{GeV}$ cut clearly has a lower efficiency in category-2. However, the background efficiency is very similar, since the lepton mis-measurement corrections take into account the differences in, e.g., p_T resolution for muons with and without SMT hits. Applying the same cuts for the different categories is therefore considered to be a reasonable choice. Optimising the p'_T cut separately for smaller subcategories would have the disadvantage of limiting the available $Z/\gamma^* \to \ell^+\ell^-$ MC statistics in the p'_T tails. Particularly in the case of category-2, the lack of MC statistics is visible as "steps" in the efficiency curves. With an inappropriate optimisation procedure, there would even be a danger of "training" the cuts against a small number of MC events and thus underestimating the background.

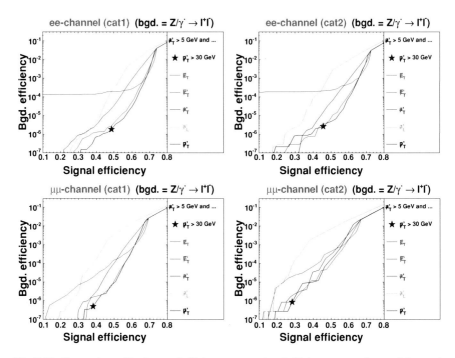

Fig. 7.21 Comparison of background efficiency versus signal efficiency curves for candidate variables that discriminate between Drell-Yan and ZZ/γ^* signal. The "soft" q'_T cut is included. The categories based on lepton p_T resolutions are defined in the text

7.8 Lepton Fake Rate Measurement

The $ZZ/\gamma^* \to \nu\bar{\nu}\ell^+\ell^-$ and $WZ/\gamma^* \to \ell\nu\ell^+\ell^-$ processes both have potential backgrounds due to the mis-reconstruction of hadronic jets as isolated electron or muon signatures. Simulation alone cannot reliably estimate these backgrounds since (i) the available event generators do not precisely predict jet rates, and (ii) the detector simulation cannot accurately describe the mis-reconstruction rates. An estimate can be made using real data with the so called "matrix method". This requires knowledge of the efficiencies for genuine (prompt) leptons (ϵ_{SIG}) and jets (ϵ_{QCD}) that satisfy certain loose requirements to also satisfy tighter requirements. The prompt lepton efficiencies are measured in a sample of $Z/\gamma^* \to \ell^+\ell^-$ candidate events using the tag-and-probe method as used in Chap. 6. A sample of events containing a high p_T jet which is back-to-back with a loose quality electron or muon is used to measure the fake rates (ϵ_{QCD}). The following requirements are imposed on this sample:

- The leading jet is matched to a trigger object at all three trigger levels, thus eliminating any trigger bias on the lepton.
- Leading jet $p_T > 20\,\text{GeV}$.

7.8 Lepton Fake Rate Measurement

- The leading jet contains at least five charged tracks, which are consistent with the primary vertex within 3 cm along the beam axis.
- The $\Delta\phi$ between the leading jet and the lepton is larger than 3 rad.
- In order to reject W+jet events, the jet + electron events must satisfy $\not{E}_T > 20\,\text{GeV}$. No $\not{E}_T$ requirement is made for jet + muon events, due to the complicated correlations between mis-reconstruction of muon p_T and isolation, and the $\not{E}_T$.

Figure 7.22 shows the ϵ_{SIG} and ϵ_{QCD} measurements as a function of the lepton p_T. Electrons in the CC, EC, and IC regions are treated separately. In order to study possible contamination of W+jet events in the QCD sample, the measurements are made in additional bins of $\Delta\phi$. Genuine dijet events are expected to be strongly peaked at $\Delta\phi \approx \pi$. Particularly for the tighter definitions, the fake rate systematically rises with decreasing $\Delta\phi$, consistent with increasing prompt lepton contamination due to W+jet events. The sample is dominated by events in the largest $\Delta\phi$ bin, and relaxing the cut (rather than looking in exclusive $\Delta\phi$ bins) gives a variation of less than 10%.

7.9 $W(+\text{jet})$ Background Estimation

An important source of background in the ZZ/γ^* analysis is W+jet production in which a jet is mis-reconstructed as an electron or muon. Since MC simulation alone cannot accurately predict this background, an estimate is made using the real data with the matrix method. The MC prediction is still used to describe the kinematic properties of the events, but is normalised to the matrix method estimate. The efficiencies for genuine leptons (ϵ_{SIG}) and jets (ϵ_{QCD}) that meet the loose requirements to also satisfy the tight requirements are measured directly from data (see Sect. 7.8).

The matrix method involves selecting a "loose" sample of events, of which the standard "tight" sample is a subset. The tight sample is based on the ZZ/γ^* candidate event selection (see Sect. 7.10), except with a relaxed invariant mass cut, to increase the sample size. Consistent cuts are made for the MC prediction in determining the correction factor to be applied to the MC events. Separate estimates are made for events passing and failing the p'_T cut (again see Sect. 7.10).

In order to take into account the p_T dependence of ϵ_{SIG} and ϵ_{QCD}, and the differences between CC, EC, and IC electrons, the two leptons must be treated separately. For example in the ee(CC+EC) category, the background from a genuine electron in the CC and a fake electron in the EC is calculated as follows. A *loose* selection allows the EC electron to be of loose quality. The number of events in this loose sample is,

$$N_L = N_L^{\text{QCD}} + N_L^{\text{SIG}}.$$

The number of events in which the EC electron also satisfies the tight requirements is

$$N_T = \epsilon_{\text{QCD}} N_L^{\text{QCD}} + \epsilon_{\text{SIG}} N_L^{\text{SIG}},$$

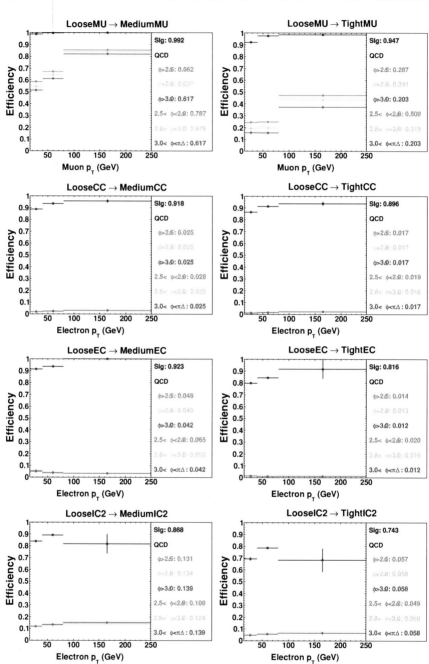

Fig. 7.22 Measurements of ϵ_{SIG} and ϵ_{QCD} for different lepton requirements (with respect to the loose requirements), as a function of the lepton p_T. The *left hand* and *right hand columns* correspond to medium and tight lepton qualities respectively

7.9 W(+jet) Background Estimation

where ϵ_{SIG} and ϵ_{QCD} correspond to EC electrons. One can then solve for N_L^{QCD},

$$N_L^{QCD} = \frac{\epsilon_{SIG} N_L - N_T}{\epsilon_{SIG} - \epsilon_{QCD}},$$

such that estimated number of W+jet events remaining in the tight sample is given by

$$N_T^{QCD} = \epsilon_{QCD} N_L^{QCD}$$

with an uncertainty (ignoring the uncertainty on the fake rates) of

$$\sigma^2(N_T^{QCD}) = N_{LNT} \left| \frac{\epsilon_{QCD} \epsilon_{SIG} N_{LNT}}{\epsilon_{SIG} - \epsilon_{QCD}} \right|^2 + N_T \left| \frac{\epsilon_{QCD}(\epsilon_{SIG} - 1) N_T}{\epsilon_{SIG} - \epsilon_{QCD}} \right|^2$$

where N_{LNT} is the number of observed loose-not-tight events. This estimate is performed in bins of the EC electron p_T. A separate estimate for a genuine electron in the EC and a fake electron in the CC, and is added to the first estimate to get the total predicted background for the CC+EC category.

For the $e\mu$ channel, there are separate estimates for $W \rightarrow \mu\nu$ (+fake e) and $W \rightarrow e\nu$ (+fake μ). In the $W \rightarrow \mu\nu$ estimate for example, the loose sample requires that the muon is tight but the electron can be loose.

Figures 7.23 and 7.24 show the predicted number of background events as a function of the lepton p_Ts (both leptons in the ee and $\mu\mu$ channels). Also shown is the prediction from MC, for the same selection cuts. The sample of MC events is scaled to the matrix method estimate in order to model the background. Unfortunately, the absolute number of MC events passing all cuts is rather limited. Therefore, we allow the MC events to contain one loose lepton, and determine the scaling factor appropriately. The right hand column of Fig. 7.23 and 7.24 show the MC predictions with the relaxed cuts.

7.10 $Z/\gamma^* \rightarrow \ell^+\ell^-$ Candidate Selection

On top of the dilepton preselection (the ZZ/γ^* analysis requires two tight quality leptons) the following requirements are made in order to select a sample of signal candidate events:

- Dilepton invariant mass between 60 and 120 GeV.
- In order to suppress $Z/\gamma^* \rightarrow \ell^+\ell^-$ and effectively eliminate $Z/\gamma^* \rightarrow \tau^+\tau^-$ altogether, we require $q'_T > 5$ GeV.
- To further suppress $Z/\gamma^* \rightarrow \ell^+\ell^-$ we require $p'_T > 30$ GeV.
- In order to eliminate events with poorly reconstructed q'_T and p'_T, we veto events with more than two reconstructed jets.

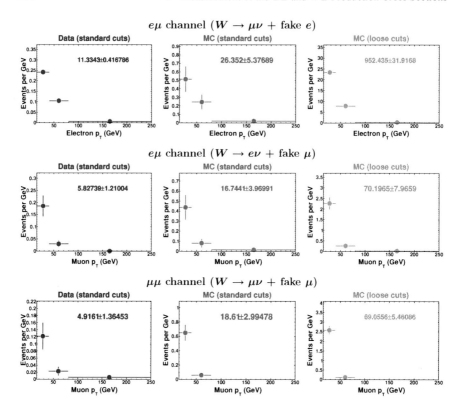

Fig. 7.23 *Left* number of background events estimated using the matrix method. *Middle* number of W+jets MC events passing the selection cuts. *Right* number of W+jets MC events passing the loose selection cuts

- In order to reject $WZ/\gamma^* \to \ell\nu\ell^+\ell^-$ and $Z/\gamma^* Z/\gamma^* \to \ell^+\ell^-\ell^+\ell^-$ there must be no additional EM clusters, muons, taus or isolated tracks meeting the respective veto quality requirements that are detailed in Sect. 7.4.

The $\not{p}'_T$ requirement is optimised by considering the predicted total uncertainty on the measured signal cross section. Figure 7.25 shows how the statistical, systematic and total uncertainties vary as a function of the cut value. Also shown are the numbers of surviving signal and background events as a function of the cut. The Drell-Yan component of the background can be seen to rapidly rise for $\not{p}'_T$ less than 25 GeV or so, having a dramatic effect on the systematic uncertainty. The predicted total uncertainty rises rather more slowly above the minimum than below the minimum. Whilst the minimum is around 25 GeV, we choose to cut at 30 GeV in order to stay safely away from the falling edge of the Drell-Yan background.

Figure 7.26 shows a number of distributions after all cuts except for the Drell-Yan rejection (q'_T and $\not{p}'_T$ cuts). This is the stage at which the normalisation is

7.10 $Z/\gamma^* \to \ell^+\ell^-$ Candidate Selection

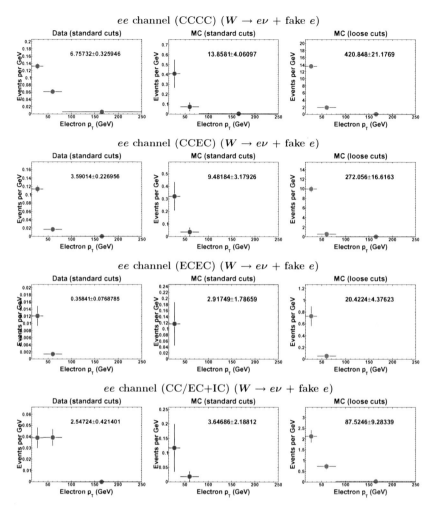

Fig. 7.24 *Left* number of background events estimated using the matrix method. *Middle* number of W+jets MC events passing the selection cuts. *Right* number of W+jets MC events passing the loose selection cuts

determined, and which will be used to measure the inclusive Z/γ^* cross section in the ratio $\sigma(ZZ/\gamma^*)/\sigma(Z/\gamma^*)$.

Figure 7.27 shows the distributions of the different cut variables. In each case, all selection cuts are made, apart from the cut on the variable in question. The $n_{\text{extra}}^{\text{lept}}$ variable is the sum of veto quality EM cluster, muon, tau and isolated track counts. The WZ/γ^* background peaks at a value of two in this distribution, since a charged lepton from a W decay is likely to be counted twice when a central track is reconstructed. Figure 7.28 shows the numbers of EM clusters, muons, taus, and isolated tracks. All selection requirements have been made, expect for the veto on the specific object

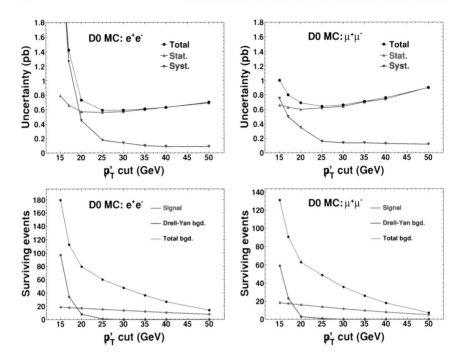

Fig. 7.25 Variation of the predicted (*top row*) signal cross section uncertainties and (*bottom row*) surviving events, with the choice of missing p_T cut

type plotted. It can be seen that the rate of additional isolated tracks and muons is somewhat underestimated by the simulation. Such a mis-modelling is mostly cancelled in the ratio of ZZ/γ^* to $Z/\gamma^* \to \ell^+\ell^-$, since the $Z/\gamma^* \to \ell^+\ell^-$ selection includes these vetoes. Tables 7.3, 7.4, and 7.5 list the predicted (broken down by process) and observed yields after all selection requirements. Also listed are the yields for the samples separately failing the mass, $\not{p}'_T$, charge, additional lepton and additional jet requirements. Systematic uncertainties (detailed in Sect. 7.14) are quoted for the predictions.

7.10.1 Signal-Free Control Regions

It is informative to look at some basic kinematic distributions in the event samples that fail each selection cut, in order to verify the background modelling. Figures 7.29, 7.30 and 7.31 show comparisons of data and simulation in various kinematic distributions. The following control regions are considered:

- **Inverted $\not{p}'_T$ requirement** The requirement, $\not{q}'_T > 5$ GeV, is still imposed so this sample is dominated by mis-reconstructed Drell-Yan events.

7.10 $Z/\gamma^* \to \ell^+\ell^-$ Candidate Selection

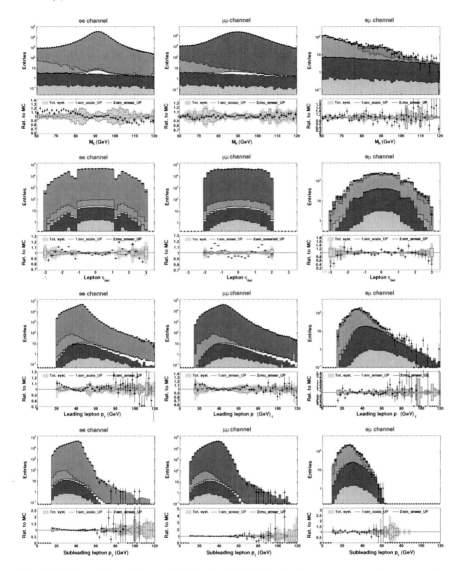

Fig. 7.26 Comparison of data and MC after all ZZ/γ^* selection cuts, except for cuts on $\not{p}'_T$ and $\not{q}'_T$

- **Inverted mass requirement** This sample is dominated by $WW \to \ell^+\nu\ell^-\bar{\nu}$ production, with a smaller contribution from W+jet and $W\gamma$ production.
- **Inverted opposite charge requirement** In order to increase statistics, the mass cut is also relaxed (i.e. only requiring $M_{\ell\ell} > 40\,\text{GeV}$). This sample is dominated by W+jet and $W\gamma$ production.

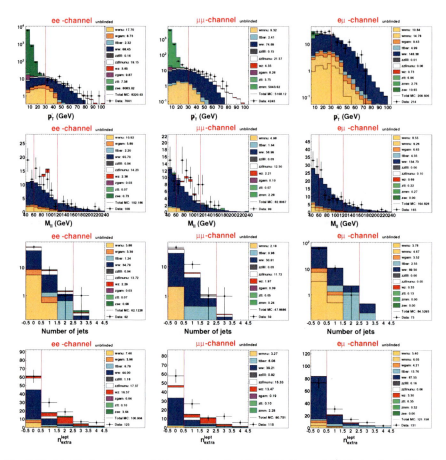

Fig. 7.27 Distribution of the variables used to select the $ZZ/\gamma^* \to \nu\bar{\nu}\ell^+\ell^-$ signal candidates. All cuts have been applied *except* for the cut on the plotted variable

- **Inverted lepton veto** This sample contains a mixture of different processes but is dominated by $WZ/\gamma^* \to \ell\nu\ell^+\ell^-$ and $t\bar{t}$.

The sample that fails the jet veto contains too few events (predicted and observed) to be interesting to plot.

7.11 $WZ/\gamma^* \to \ell\nu\ell^+\ell^-$ Signal Selection

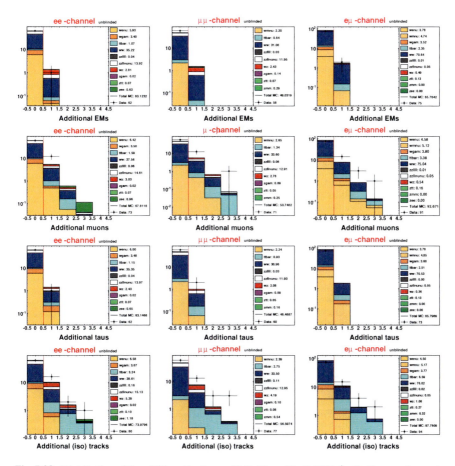

Fig. 7.28 Distribution of the additional lepton multiplicities with all ZZ/γ^* selection requirements imposed (including the other additional activity vetoes) except for the cut on the plotted variable

7.11 $WZ/\gamma^* \to \ell\nu\ell^+\ell^-$ Signal Selection

7.11.1 Selection Cuts

The basic signature of $WZ/\gamma^* \to \ell\nu\ell^+\ell^-$ is three high p_T charged leptons, two of which are of the same flavour and opposite charge, produced in association with missing transverse momentum due to the neutrino. The dominant backgrounds arise from $Z/\gamma^* \to \ell^+\ell^-$ production in association with a photon or jet that is misreconstructed as an electron or muon. This mis-reconstruction can also contribute to a fake $\displaystyle{\not}E_T$. Since the signal and background both involve a $Z/\gamma^* \to \ell^+\ell^-$ decay, we can potentially improve the resolution of the $\displaystyle{\not}E_T$ by constraining the $Z/\gamma^* \to \ell^+\ell^-$ invariant mass to m_Z. The p_Ts of the leptons that are assigned to be Z/γ^*

Table 7.3 Table of predicted signal and background yields for the $ZZ/\gamma^* \to \nu\bar{\nu}e^+e^-$ signal and control regions

Process	Accepted	Rejected by cut on p'_T	M_{ll}	Extra lep.	$\sum q$	n_{jet}
$Z/\gamma^* \to e^+e^-$	0.6 ± 0.5	11986.7 ± 2417.0	0.2 ± 7.6	3.4 ± 2.6	0.0 ± 0.0	0.1 ± 0.1
$Z/\gamma^* \to \tau^+\tau^-$	0.1 ± 0.1	8.6 ± 3.5	0.0 ± 0.7	0.0 ± 0.1	0.0 ± 0.0	0.0 ± 0.0
$WW \to \ell^+\nu\ell^-\bar{\nu}$	34.7 ± 1.4	35.5 ± 1.2	33.0 ± 1.4	9.3 ± 16.2	0.3 ± 0.1	0.1 ± 0.1
$WZ/\gamma^* \to \ell\nu\ell^+\ell^-$	2.3 ± 0.1	1.9 ± 0.1	0.1 ± 0.0	14.2 ± 5.9	0.2 ± 0.0	0.0 ± 0.0
$W \to e\nu$	5.9 ± 2.3	13.0 ± 5.1	5.1 ± 2.4	1.6 ± 5.0	4.4 ± 2.2	0.0 ± 0.0
$W\gamma \to e\nu\gamma$	3.4 ± 0.4	5.4 ± 0.4	2.8 ± 0.6	0.6 ± 1.1	3.3 ± 0.4	0.0 ± 0.0
$Z/\gamma^* Z/\gamma^* \to \ell^+\ell^-\ell^+\ell^-$	0.0 ± 0.0	0.1 ± 0.0	0.0 ± 0.0	1.3 ± 0.5	0.0 ± 0.0	0.0 ± 0.0
$t\bar{t}$	1.0 ± 0.2	1.5 ± 0.2	1.2 ± 0.2	6.6 ± 2.9	0.0 ± 0.0	0.2 ± 0.1
Total background	48.0 ± 2.7	12070.4 ± 2418.4	42.4 ± 9.5	37.0 ± 33.1	8.2 ± 2.4	0.4 ± 0.2
$ZZ/\gamma^* \to \nu\bar{\nu}\ell^+\ell^-$	13.7 ± 0.4	7.4 ± 0.2	0.6 ± 0.0	3.7 ± 6.3	0.2 ± 0.0	0.1 ± 0.0
Predicted total	61.7 ± 2.8	12077.8 ± 2418.4	43.1 ± 9.5	40.8 ± 39.3	8.4 ± 2.4	0.4 ± 0.3
Observed	61	10560	50	63	12	1

The quoted uncertainties on the predictions are systematic

Table 7.4 Table of predicted signal and background yields for the $ZZ/\gamma^* \to \nu\bar{\nu}\mu^+\mu^-$ signal and control regions

Process	Accepted	Rejected by cut on p'_T	M_{ll}	Extra lep.	$\sum q$	n_{jet}
$Z/\gamma^* \to \mu^+\mu^-$	0.2 ± 0.7	8459.4 ± 1480.1	3.2 ± 2.5	2.2 ± 2.1	0.4 ± 0.3	0.1 ± 0.2
$Z/\gamma^* \to \tau^+\tau^-$	0.0 ± 0.0	5.0 ± 2.3	0.0 ± 0.0	0.1 ± 0.0	0.0 ± 0.0	0.0 ± 0.0
$WW \to \ell^+\nu\ell^-\bar{\nu}$	30.5 ± 1.4	7.4 ± 2.6	28.9 ± 1.4	8.7 ± 13.7	0.0 ± 0.0	0.1 ± 0.1
$WZ/\gamma^* \to \ell\nu\ell^+\ell^-$	2.0 ± 0.1	2.7 ± 0.2	0.3 ± 0.0	11.7 ± 4.9	0.2 ± 0.0	0.0 ± 0.0
$W \to \mu\nu$	2.2 ± 1.1	9.6 ± 3.0	2.8 ± 1.1	1.1 ± 1.5	0.7 ± 0.3	0.0 ± 0.0
$Z/\gamma^* Z/\gamma^* \to \ell^+\ell^-\ell^+\ell^-$	0.0 ± 0.0	0.2 ± 0.0	0.0 ± 0.0	0.9 ± 0.3	0.0 ± 0.0	0.0 ± 0.0
$t\bar{t}$	0.8 ± 0.1	2.1 ± 0.2	0.8 ± 0.1	6.2 ± 2.7	0.0 ± 0.0	0.2 ± 0.1
Total background	35.8 ± 2.4	8542.0 ± 1482.4	36.1 ± 2.7	30.9 ± 24.3	1.3 ± 0.4	0.4 ± 0.3
$ZZ/\gamma^* \to \nu\bar{\nu}\ell^+\ell^-$	11.7 ± 0.4	10.9 ± 0.4	0.8 ± 0.1	3.7 ± 5.6	0.0 ± 0.0	0.0 ± 0.0
Predicted total	47.5 ± 2.5	8552.9 ± 1482.5	36.9 ± 2.7	34.6 ± 29.8	1.3 ± 0.4	0.4 ± 0.3
Observed	58	7416	42	60	4	1

The quoted uncertainties on the predictions are systematic

daughters are floated within 3 standard deviations of their resolution $\sigma(p_T)$ in a fit that minimises the following χ^2 function:

$$\chi^2 = \left(\frac{M_{\ell\ell} - m_Z}{\Gamma_Z}\right)^2 + \left(\frac{\delta p_T^{(1)}}{\sigma(p_T^{(1)})}\right)^2 + \left(\frac{\delta p_T^{(2)}}{\sigma(p_T^{(2)})}\right)^2,$$

7.11 $WZ/\gamma^* \to \ell\nu\ell^+\ell^-$ Signal Selection

Table 7.5 Table of predicted yields in the $e\mu$ channel

Process	Accepted	Rejected by cut on				
		$\not{E}'_T$	M_{ll}	Extra lep.	$\sum q$	n_{jet}
$Z/\gamma^* \to e^+e^-$	0.0 ± 0.0	15.8 ± 8.0	0.0 ± 0.0	0.0 ± 0.0	0.0 ± 0.0	0.0 ± 0.0
$Z/\gamma^* \to \mu^+\mu^-$	0.0 ± 0.0	5.9 ± 2.5	0.3 ± 0.5	0.3 ± 0.5	0.0 ± 0.0	0.0 ± 0.0
$Z/\gamma^* \to \tau^+\tau^-$	0.1 ± 0.1	14.9 ± 6.0	0.1 ± 0.9	0.2 ± 0.2	0.0 ± 0.0	0.0 ± 0.0
$WW \to \ell^+\nu\ell^-\bar{\nu}$	69.2 ± 2.9	84.5 ± 3.0	67.6 ± 3.0	18.4 ± 31.3	0.4 ± 0.1	0.3 ± 0.2
$WZ/\gamma^* \to \ell\nu\ell^+\ell^-$	0.3 ± 0.0	0.4 ± 0.1	0.4 ± 0.1	3.0 ± 1.2	0.3 ± 0.0	0.0 ± 0.0
$W \to e\nu$	3.8 ± 2.1	8.1 ± 5.0	4.9 ± 2.2	1.6 ± 2.7	1.2 ± 0.8	0.0 ± 0.0
$W \to \mu\nu$	4.7 ± 4.0	11.9 ± 9.6	4.6 ± 4.3	1.4 ± 3.5	2.8 ± 2.5	0.0 ± 0.0
$W\gamma \to e\nu\gamma$	3.5 ± 0.7	6.3 ± 0.9	3.4 ± 0.3	0.7 ± 1.7	3.2 ± 0.4	0.0 ± 0.0
$Z/\gamma^*Z/\gamma^* \to \ell^+\ell^-\ell^+\ell^-$	0.0 ± 0.0	0.0 ± 0.0	0.0 ± 0.0	0.2 ± 0.1	0.0 ± 0.0	0.0 ± 0.0
$t\bar{t}$	2.3 ± 0.2	3.3 ± 0.3	2.1 ± 0.3	13.3 ± 5.8	0.0 ± 0.0	0.3 ± 0.1
Total background	83.9 ± 6.0	151.2 ± 19.2	83.3 ± 6.8	39.1 ± 46.0	7.9 ± 3.1	0.6 ± 0.3
$ZZ/\gamma^* \to \nu\bar{\nu}\ell^+\ell^-$	0.0 ± 0.0	0.0 ± 0.0	0.1 ± 0.0	0.0 ± 0.0	0.0 ± 0.0	0.0 ± 0.0
Predicted total	84.0 ± 6.0	151.2 ± 19.2	83.3 ± 6.8	39.1 ± 46.0	7.9 ± 3.1	0.6 ± 0.3
Observed	73	162	96	60	8	0

The quoted uncertainties on the predictions are systematic

where m_Z and Γ_Z are the masses and widths of the Z respectively [12]. The uncertainties on the lepton p_Ts are estimated in the same way as for the construction of the missing p_T estimators used in the ZZ/γ^* analysis (see Sect. 7.7). Figure 7.32 compares the Z/γ^*+jet and $Z/\gamma^*+\gamma$ background rejection versus signal efficiency curves achieved by the standard and constrained $\not{E}_T$. A cut at 20 GeV cut is indicated by the star shaped marker. The kinematic constraint makes a small improvement in most of the sub-channels, particularly those with a $Z/\gamma^* \to \mu^+\mu^-$ decay.

The WZ/γ^* dilepton preselection requires two medium quality leptons as described in Sect. 7.3. Figures 7.33, 7.34, 7.35 and 7.36 compare data with simulation at the dilepton preselection level in a number of kinematic distributions. The excess at lower values of $M_{\ell\ell}$ in the ee channel is likely to be a missing mulitjet background component. This is not considered to be an important background after making the WZ/γ^* candidate selection requirements. Also, this level of background will have an effect on the normalisation sample that is negligible compared to the signal cross section measurement uncertainties. Figure 7.37 shows the increased dilepton yield in data for medium-medium requirements compared to the tight–tight requirements in the ZZ/γ^* selection.

The following additional requirements are made on top of the dilepton preselection in order to select a sample of WZ/γ^* signal candidate events:

- $60 < M_{\ell\ell} < 120$ GeV,
- leading lepton $p_T > 25$ GeV,
- A third tight CC/EC electron or tight muon with $p_T > 15$ GeV. IC electrons are not considered since they are more easily faked by hadronic jets. If there are three leptons of the same flavour, then there are may be two possible opposite sign

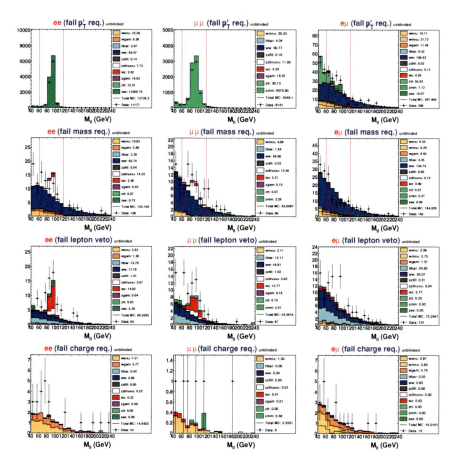

Fig. 7.29 Data-vs-MC comparisons of the M_{ll} distribution in the different ZZ/γ^* signal-free control regions

combinations. In this case, the pair with smallest $|M_{\ell\ell} - m_Z|$, are assigned to the $Z/\gamma^* \to \ell^+\ell^-$ decay. Events are rejected if the assigned W daughter fails the tight requirement, even if one of the Z/γ^* daughters is of tight quality.

- Constrained $\slashed{E}_T > 20\,\text{GeV}$.
- If the third lepton is an electron then we require $|M_{\ell\ell\ell} - 91.2| > |M_{\ell\ell} - 91.2|$. This requirement rejects events in which a wide-angle photon is emitted from a lepton in a $Z/\gamma^* \to \ell^+\ell^-$ decay.
- No additional veto-quality EM clusters for the sub-channels containing a $W \to e\nu$ decay, and no additional veto-quality muons for the sub-channels containing a $W \to \mu\nu$ decay. The requirements on the veto-quality leptons are detailed in Sect. 7.4.

7.11 $WZ/\gamma^* \to \ell\nu\ell^+\ell^-$ Signal Selection

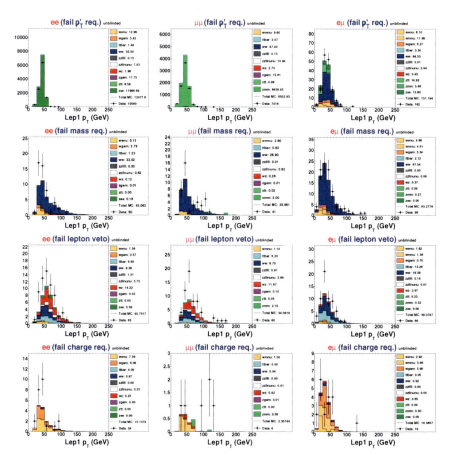

Fig. 7.30 Data-vs-MC comparisons of the leading lepton p_T distribution in the different ZZ/γ^* signal-free control regions

Figure 7.38 shows the dilepton invariant mass and $\not{E}_T$ distributions after applying all WZ/γ^* selection requirements except for the cut on the plotted variable. Also shown are the numbers of additional veto quality muons or electrons, depending on the sub-channel. Figure 7.39 shows the two dimensional projections in $\not{E}_T$ and transverse mass for signal and background. The transverse mass is defined as

$$M_T = \sqrt{2 p_T^{(3)} \not{E}_T (1 - \cos \Delta\phi)},$$

where $p_T^{(3)}$ is the transverse momentum of the third lepton, and $\Delta\phi$ is the difference in azimuth between the this lepton and the $\not{E}_T$. Figure 7.40 shows the dilepton p_T, W p_T, and M_T distributions for the signal candidates. Figure 7.41 shows the p_T

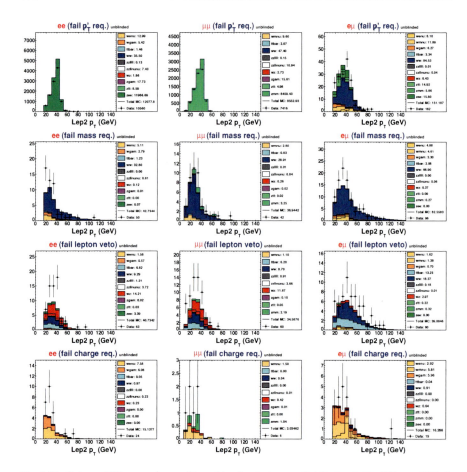

Fig. 7.31 Data-vs-MC comparisons of the subleading lepton p_T distribution in the different ZZ/γ^* signal-free control regions

distributions separately for the leading and sub-leading Z/γ^* daughter leptons, and the W daughter lepton.

Figure 7.42 shows the fraction of MC events in which the leptons are correctly assigned to the Z/γ^* and W (only in the eee and $\mu\mu\mu$ channels which have potential ambiguities). The mis-assignment rate is $\approx 5\,\%$ in the eee channel and $\approx 13\,\%$ in the $\mu\mu\mu$ channel. The larger rate in the $\mu\mu\mu$ channel is easily traced to the relatively poor $M_{\ell\ell}$ resolution. Also shown are the $M_{\ell\ell}$ and m_T distributions for the events with correct and incorrect assignment. Since the choice of association is based on the $M_{\ell\ell}$, there is little difference in shape for the two correct and incorrect samples. The m_T distribution is however significantly broader in the incorrect sample.

7.11 $WZ/\gamma^* \to \ell\nu\ell^+\ell^-$ Signal Selection

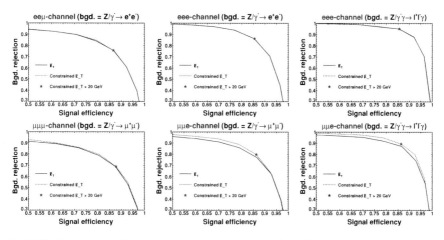

Fig. 7.32 Comparison of background rejection versus signal efficiency *curves* for the standard (*solid black*) and constrained (*dashed red*) $\displaystyle{\not}E_T$ variables. The different subfigures correspond to different sub-channels and either $Z/\gamma^* \to \ell^+\ell^-$ or $Z/\gamma^*\gamma \to \ell^+\ell^-\gamma$ as the background

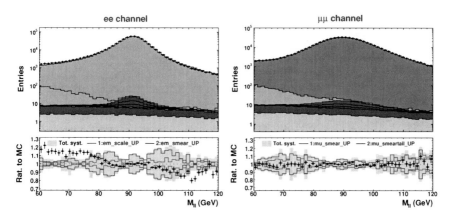

Fig. 7.33 Data-vs-MC comparison of the $M_{\ell\ell}$ distribution at the (WZ/γ^*) dilepton stage

Tables 7.6 and 7.7 list the observed yields and predicted signal and background yields for each sub-channel. The number of events failing the $\displaystyle{\not}E_T$ cut is also provided. The quoted uncertainties are systematic (more details in Sect. 7.14).

7.11.2 Normalisation of Z/γ^*+jet Backgrounds

The simulated $Z/\gamma^* \to \ell^+\ell^-$ events are not expected to accurately model the probability for jets to fake isolated electron or muon signatures. Section 7.9 discusses this problem in the context of estimating W+jets backgrounds for the ZZ/γ^* analysis. We apply a similar approach for modelling the Z+jet backgrounds here. The number

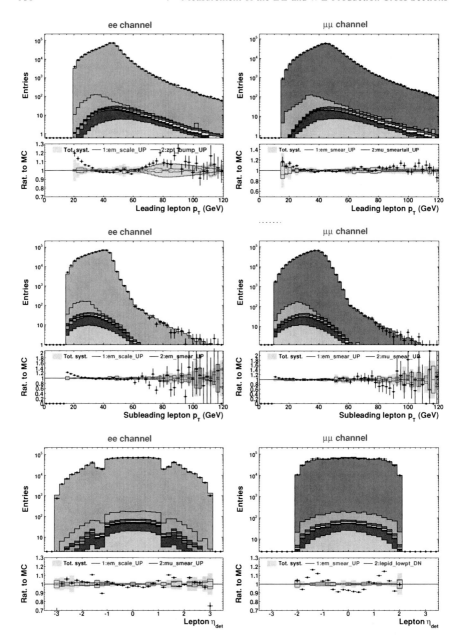

Fig. 7.34 Data-vs-MC comparison at the (WZ/γ^*) dilepton stage, of the p_T distributions of the (*top row*) leading lepton and (*middle row*) sub-leading lepton. The *bottom row* shows the lepton η_{det} distribution (two entries per event)

7.11 $WZ/\gamma^* \to \ell\nu\ell^+\ell^-$ Signal Selection

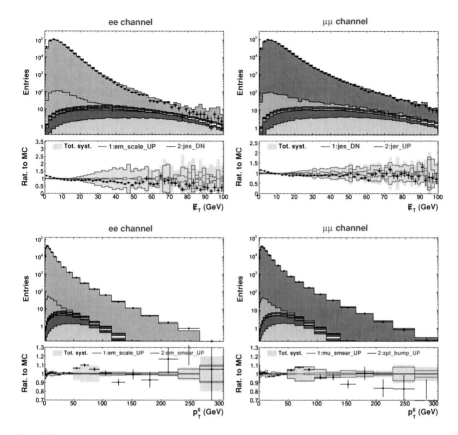

Fig. 7.35 Data-vs-MC comparisons of the (*top*) $\displaystyle{\not}E_T$ and (*bottom*) dilepton p_T distributions at the (WZ/γ^*) dilepton stage

of background events is estimated from data using the matrix method, and a sample of Z/γ^* MC events is scaled to match this estimate.

A loose sample relaxes the third lepton requirement from tightquality to loosequality. In order to increase the sample size, the $\displaystyle{\not}E_T$ requirement is also relaxed. The matrix method evaluates the number of expected background events as a function of the third lepton p_T, taking into account the p_T dependence of ϵ_{SIG} and ϵ_{QCD}. Events with a third CC electron and a third EC electron are treated separately. Figures 7.43 and 7.44 show for the *ee* and $\mu\mu$ channels respectively, the estimated number of background events from the matrix method, and the prediction from MC. We have sufficient $Z/\gamma^* \to \ell^+\ell^-$ MC statistics such that relaxing the quality cuts is not necessary (unlike the simulation of W+jets in the ZZ/γ^* analysis).

182   7 Measurement of the ZZ and WZ Production Cross Sections

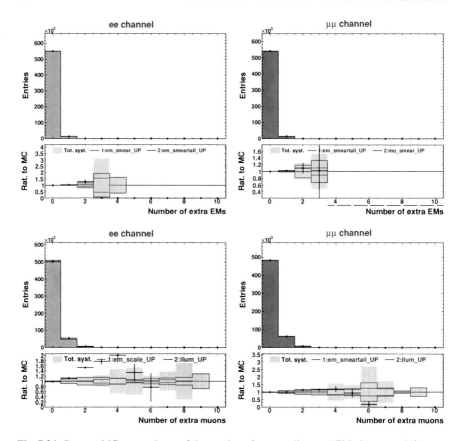

Fig. 7.36 Data-vs-MC comparisons of the number of veto quality (*top*) EM clusters and (*bottom*) muons

7.11.3 Trigger Efficiencies

The WZ/γ^* (and also ZZ/γ^*) cross sections are measured as a ratio to the Z/γ^* cross section. No specific trigger requirement is made in either analysis. We refer to this as "inclusive" triggering. Whilst the ZZ/γ^* analysis is expected to be insensitive to trigger biases, the WZ/γ^* selection requires an additional high p_T lepton compared to the inclusive dilepton selection, which is likely to increase the trigger efficiency. This may have a significant effect in the $Z/\gamma^* \to \mu^+\mu^-$ channels, since the single muon triggers are roughly 60–70 % efficient (per muon). Figure 7.45 shows the single muon trigger efficiencies with respect to medium quality offline muons in the inclusive dilepton sample, measured using the tag-and-probe method described in Chap. 6. Figure 7.46 shows the dimuon invariant mass distribution for each of the following:

- Data with the requirement of single-muon trigger.

7.11 $WZ/\gamma^* \to \ell\nu\ell^+\ell^-$ Signal Selection

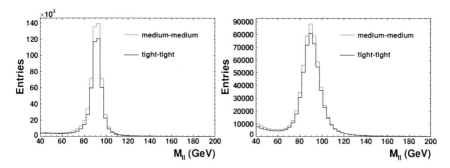

Fig. 7.37 Comparison of the dilepton yields in data for tight–tight and medium–medium requirements

Table 7.6 Table of predicted and observed yields in the two $Z/\gamma^* \to e^+e^-$ sub-channels

	$e^+e^-e^\pm$ Accepted	$e^+e^-e^\pm$ Rejected	$e^+e^-\mu^\pm$ Accepted	$e^+e^-\mu^\pm$ Rejected
$Z/\gamma^* \to \ell^+\ell^-$	0.37 ± 0.22	8.68 ± 1.39	3.24 ± 1.36	6.60 ± 1.75
$Z/\gamma^*\gamma \to l^+l^-\gamma$	0.56 ± 0.21	10.10 ± 0.46	0.07 ± 0.03	0.10 ± 0.04
$Z/\gamma^*Z/\gamma^* \to \ell^+\ell^-\ell^+\ell^-$	0.58 ± 0.10	1.05 ± 0.10	1.47 ± 0.04	0.68 ± 0.06
$t\bar{t}$	0.03 ± 0.01	0.00 ± 0.01	0.03 ± 0.01	0.01 ± 0.01
$ZZ/\gamma^* \to \nu\bar{\nu}\ell^+\ell^-$	0.00 ± 0.00	0.00 ± 0.00	0.00 ± 0.00	0.00 ± 0.00
$WW \to \ell^+\nu\ell^-\bar{\nu}$	0.03 ± 0.01	0.00 ± 0.00	0.00 ± 0.01	0.01 ± 0.01
Predicted background	1.56 ± 0.44	19.83 ± 1.49	4.82 ± 1.37	7.40 ± 1.76
$WZ/\gamma^* \to \ell\nu\ell^+\ell^-$	9.83 ± 0.15	1.59 ± 0.06	13.85 ± 0.41	2.16 ± 0.07
Predicted total	11.39 ± 0.45	21.42 ± 1.49	18.67 ± 1.44	9.56 ± 1.77
Observed	17	32	17	6

The quoted uncertainties are systematic

- Data with the requirement of a single-muon OR dimuon trigger.
- Data with inclusive triggering.
- MC having simulated the requirement of at least one muon firing a single muon trigger. This distribution has been normalised to have the same integral as the data with single-muon triggers.
- MC without any trigger simulation, but scaled by exactly the same factor as in the previous bullet.

The efficiency in data can now be estimated as 82% for single-muon triggers, 87% for an OR of single-muon and dimuon triggers, and 93% for the inclusive trigger. Assuming an efficiency (per lepton) of $\approx 70\%$ for the single muon triggers, and $\approx 90\%$ for the single electron triggers would yield efficiencies of ≈ 97 and $\approx 99\%$ in the $\mu\mu + \mu$ and $\mu\mu + e$ channels respectively. The triggers therefore introduce a bias of roughly 5% in the ratio of WZ/γ^* and $Z/\gamma^* \to \ell^+\ell^-$ acceptances.

Muon+jet triggers are responsible for most of the additional efficiency of the inclusive trigger compared to the OR of single-muon and dimuon triggers. Figure 7.47 shows, as a function of the jet multiplicity and the p_T sum of the leading two jets,

Table 7.7 Table of predicted and observed yields in the two $Z/\gamma^* \to \mu^+\mu^-$ sub-channels

	$\mu^+\mu^-e^\pm$		$\mu^+\mu^-\mu^\pm$	
	Accepted	Rejected	Accepted	Rejected
$Z/\gamma^* \to \ell^+\ell^-$	1.42 ± 0.66	11.99 ± 2.08	3.91 ± 1.96	3.09 ± 1.04
$Z/\gamma^*\gamma \to l^+l^-\gamma$	1.62 ± 0.36	12.95 ± 0.58	0.09 ± 0.03	0.08 ± 0.03
$Z/\gamma^*Z/\gamma^* \to \ell^+\ell^-\ell^+\ell^-$	0.91 ± 0.15	1.45 ± 0.15	1.56 ± 0.05	0.67 ± 0.04
$t\bar{t}$	0.27 ± 0.01	0.04 ± 0.02	0.13 ± 0.02	0.02 ± 0.01
$ZZ/\gamma^* \to \nu\bar{\nu}\ell^+\ell^-$	0.00 ± 0.00	0.00 ± 0.00	0.00 ± 0.00	0.00 ± 0.00
$WW \to \ell^+\nu\ell^-\bar{\nu}$	0.01 ± 0.00	0.00 ± 0.00	0.00 ± 0.00	0.00 ± 0.00
Predicted background	4.23 ± 0.91	26.44 ± 2.17	5.70 ± 1.95	3.85 ± 1.05
$WZ/\gamma^* \to \ell\nu\ell^+\ell^-$	13.91 ± 0.29	2.12 ± 0.12	14.77 ± 0.38	2.01 ± 0.08
Predicted total	18.14 ± 0.94	28.56 ± 2.15	20.46 ± 1.96	5.86 ± 1.07
Observed	26	23	25	12

The quoted uncertainties are systematic

the ratio of the dimuon yield in data with no trigger requirement to the yield with the requirement of a single-muon or dimuon trigger. This ratio can be seen to rise systematically with increasing jet activity.

The trigger is simulated as follows: Dimuon events are assigned a weight of 0.97 if there is a reconstructed third lepton (electron or muon) with at least loose quality and $p_T > 15\,\text{GeV}$. Otherwise, they are assigned a weight of 0.87 to simulate the single-muon and dimuon triggers. A parameterisation of the ratio shown in Fig. 7.47 as a function of the jet p_T sum is used to simulate the additional efficiency of the muon+jet triggers. An additional weight factor varies between around 1.05 and a maximum of 1.15.

A grossly over conservative uncertainty of 50% on this correction results in an uncertainty of roughly 1% on the signal cross section after combining the four sub-channels. Thus, compared to an overall uncertainty of roughly 15%, we consider the uncertainty due to triggers to be negligible after making the above correction.[4]

7.11.4 Signal-Free Control Regions

Two signal-free control regions are defined for the WZ/γ^* analysis:

- **Inverted $\slashed{E}_T$ requirement** The region $\slashed{E}_T < 15\,\text{GeV}$ is dominated by Z(+jets) and $Z\gamma$. All other cuts are exactly the same as the signal selection. Figures 7.48 and 7.49 compare data and MC in this control region in various distributions.
- **Inverted charge requirement** This sample contains events that have no same flavour opposite charge sign lepton pair but otherwise pass all $WZ/\gamma^* \to \ell\nu\ell^+\ell^-$

---

[4] An alternative approach would have been to require that one of the muons assigned as a Z/γ^* daughter is matched to a single muon trigger, thus eliminating trigger bias altogether (to a good approximation at least). This would result in a ≈20% reduction in signal acceptance in the $Z/\gamma^* \to \mu^+\mu^-$ channels. Given that correcting the bias does not introduce any significant uncertainty, the inclusive trigger approach is clearly the better option.

7.11 $WZ/\gamma^* \to \ell\nu\ell^+\ell^-$ Signal Selection

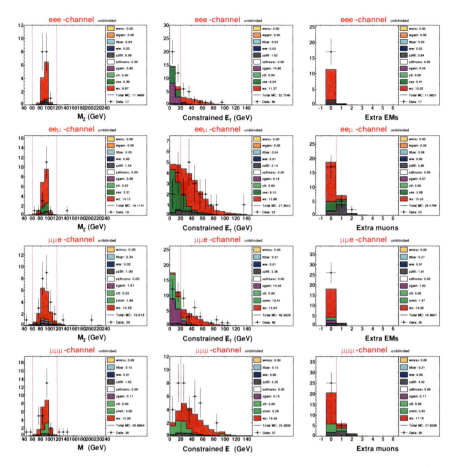

Fig. 7.38 Selection variables *before their respective cuts*, but with all other $WZ/\gamma^* \to \ell\nu\ell^+\ell^-$ cuts applied. The *rows* correspond to different sub-channels

selection cuts. Figure 7.50 compares data and MC in this sample. A small number of events are observed in data, but are consistent with MC predictions.

7.12 $ZZ/\gamma^* \to \nu\bar{\nu}\ell^+\ell^-$ Multivariate Analysis

The sample of selected $ZZ/\gamma^* \to \nu\bar{\nu}\ell^+\ell^-$ candidates is dominated by background from $WW \to \ell^+\nu\ell^-\bar{\nu}$. In discriminating signal from background, the single most powerful kinematic variable is the dilepton invariant mass as shown in Fig. 7.51.

The previous D0 analysis of this channel combined the dilepton invariant mass with a number of other kinematic variables in a simple likelihood discriminant. We attempt to further improve the separation by using a more advanced multivariate

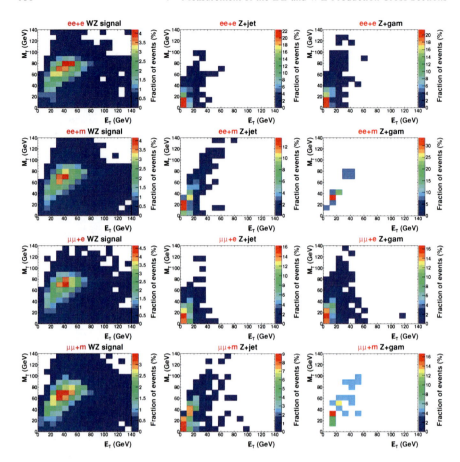

Fig. 7.39 Distribution of $\slashed{E}_T$ versus M_T for (*left*) signal, (*middle*) Z+jets background and (*right*) $Z/\gamma^* + \gamma$ background. The *rows* correspond to different sub-channels

classifier within the TMVA framework [13]. Figures 7.52 (*ee* channel) and 7.53 ($\mu\mu$ channel) compare the shapes of the input variables for signal and background. Here, the signal is $ZZ/\gamma^* \to \nu\bar{\nu}\ell^+\ell^-$ and background is *only* $WW \to \ell^+\nu\ell^-\bar{\nu}$. We choose not to train against the next largest background (W+jet) due to the limited statistics. The lower halves Figs. 7.52 and 7.53 show the input variable correlation matrices separately for signal and background. The input variables are as follows (with discussion referring to Figs. 7.52 and 7.53):

- $\sigma(M_{\ell\ell} - M_Z)$ This variable is defined as $|M_{\ell\ell} - 91.2|/\sigma(M_{\ell\ell})$, where $\sigma(M_{\ell\ell})$ is the estimated invariant mass resolution based on the lepton momentum transfer functions.
- $\slashed{E}_T$ This variable is the standard calorimeter $\slashed{E}_T$ defined in Sect. 7.4. The $\slashed{E}_T$ tends to be larger for the signal, since the two neutrinos tend to be more collimated in the signal.

7.12 $ZZ/\gamma^* \to \nu\bar{\nu}\ell^+\ell^-$ Multivariate Analysis

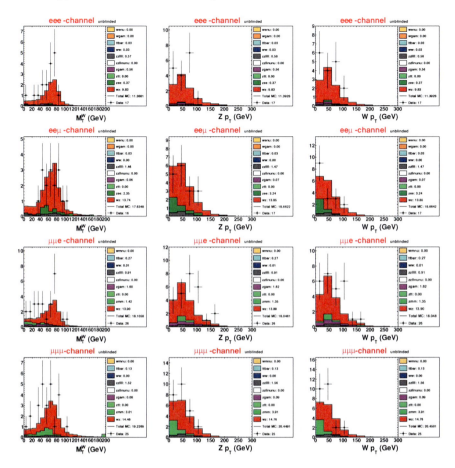

Fig. 7.40 Distributions of the $WZ/\gamma^* \to \ell\nu\ell^+\ell^-$ candidates after all cuts. The *rows* correspond to different sub-channels

- **cos θ^*** The scattering angle of the negatively charged lepton in the approximate dilepton rest frame, defined as $\cos\theta^* = \tanh[(\eta^{(-)} - \eta^{(+)})/2]$ as introduced in Chap. 5. The background tends to have a lower value of $\cos\theta^*$.
- **$\Delta\phi(l1, ll)$** The azimuthal opening angle between the leading lepton and the dilepton system. The background tends to have a larger value of $\Delta\phi(l1, ll)$.
- **Lep1/2 p_T** The background tends to have softer lepton p_T distributions; in particular for the leading lepton.

Figure 7.54 shows the signal efficiency vs. background rejection for three possible algorithms; a neural network (MLP), a boosted decision tree (BDT) and a simple likelihood. Of the three methods, the MLP gives the highest performance. Also shown in Fig. 7.54 is a comparison of the training/testing and signal/background outputs for the MLP method.

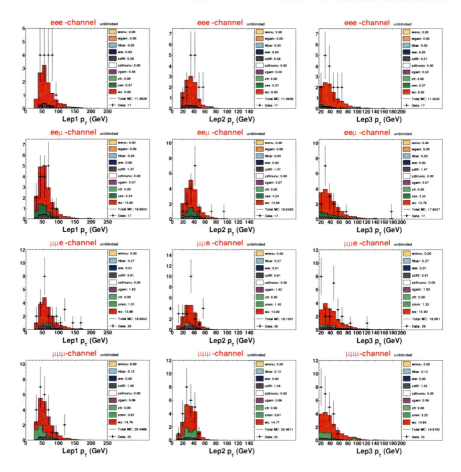

Fig. 7.41 Distributions of the $WZ/\gamma^* \to \ell\nu\ell^+\ell^-$ candidates after all cuts. The rows correspond to different sub-channels

Figure 7.55 compares data and simulation for each of the input variables at the ZZ/γ^* candidate stage. The $e\mu$ channel is particularly useful here in verifying that we accurately describe the kinematic properties of the background.

The top row of Fig. 7.56 shows the MLP output distribution of the signal candidates in the ee and $\mu\mu$ channels. The remaining rows show the MLP output distributions in the signal-free control regions. Figure 7.57 shows the equivalent distributions in the $e\mu$ channel. Since there is no $ZZ/\gamma^* \to \nu\bar{\nu}\ell^+\ell^-$ signal to train on in the $e\mu$ channel, the outputs are shown for both the ee and $\mu\mu$ versions of the MLP.

7.13 Signal Cross Section Measurements

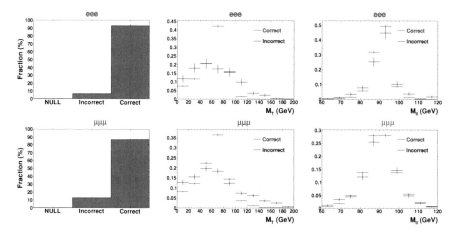

Fig. 7.42 The *left-hand column* shows the fraction of signal events in which the three leptons are correctly assigned to the Z/γ^* and W. The *middle* and *right-hand columns* show the shapes of the $M_{\ell\ell}$ and M_T distributions for events with correct and incorrect assignment

Table 7.8 Table of inclusive dilepton (A_{Z/γ^*}) and signal (A_{sig}) acceptances for the different sub-channels

Sub-channel	A_{Z/γ^*}	A_{sig}
$WZ/\gamma^* \to e\nu e^+ e^-$	0.248 ± 0.006	0.107 ± 0.003
$WZ/\gamma^* \to \mu\nu e^+ e^-$	0.248 ± 0.006	0.150 ± 0.006
$WZ/\gamma^* \to e\nu \mu^+ \mu^-$	0.268 ± 0.010	0.153 ± 0.007
$WZ/\gamma^* \to \mu\nu \mu^+ \mu^-$	0.268 ± 0.010	0.160 ± 0.009
$ZZ/\gamma^* \to \nu\bar{\nu} e^+ e^-$	0.178 ± 0.007	0.109 ± 0.006
$ZZ/\gamma^* \to \nu\bar{\nu} \mu^+ \mu^-$	0.186 ± 0.011	0.090 ± 0.006

The quoted uncertainties are systematic

7.13 Signal Cross Section Measurements

The signal production cross sections are determined as a ratio to the inclusive Z/γ^* cross section as follows:

$$\sigma_{\text{sig}} = \sigma_{Z/\gamma^*} \times \frac{A_{Z/\gamma^*}}{A_{\text{sig}}} \times \frac{N_{\text{sig}}^{\text{obs}}}{N_{Z/\gamma^*}^{\text{obs}}}$$

where σ_{Z/γ^*} is the calculated inclusive Z/γ^* cross section; A_{Z/γ^*} and A_{sig} are the acceptances (multiplied by efficiencies) for Z/γ^* and signal respectively; $N_{Z/\gamma^*}^{\text{obs}}$ and $N_{\text{sig}}^{\text{obs}}$ are the observed numbers of (background subtracted) Z/γ^* and signal events respectively. Table 7.8 lists the values of A_{Z/γ^*} and A_{sig} for the different sub-channels. The cross section times branching fraction for $Z/\gamma^* \to l^+l^-$ (one lepton flavour) has been calculated in [14] using a modified version of the code by

190　　　　　　　　　　7　Measurement of the ZZ and WZ Production Cross Sections

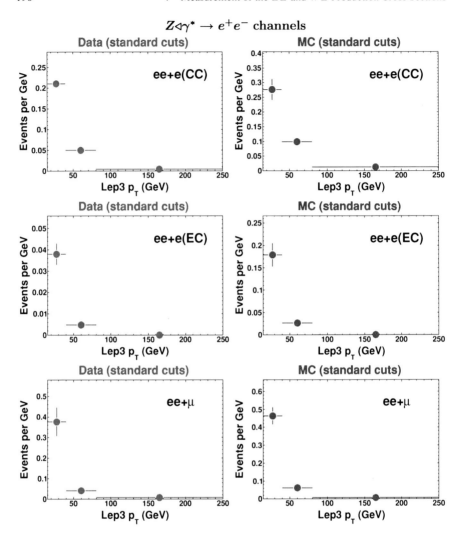

Fig. 7.43 *Left* matrix method estimate of the Z+jet background. *Right* MC prediction

Hamburg, Matsura and van Neerven [15] with the MRST2004 NNLO PDFs [16]. This code explicitly excludes the γ^* and Z/γ^* interference. Therefore, a correction of 1.0186 ± 0.0007 was determined using MC@NLO and PYTHIA The final value is

$$\sigma(p\bar{p} \to Z/\gamma^*) \times B(Z/\gamma^* \to \ell^+\ell^-) = 256.6^{+5.1}_{-12.0} \text{ pb},$$

with $60 < M_{\ell\ell} < 130$ GeV.

We can safely ignore backgrounds for the Z/γ^* yield, and $N^{\text{obs}}_{Z/\gamma^*}$ is simply the number of observed events at the appropriate normalisation stage (see Sects. 7.10 and

7.13 Signal Cross Section Measurements

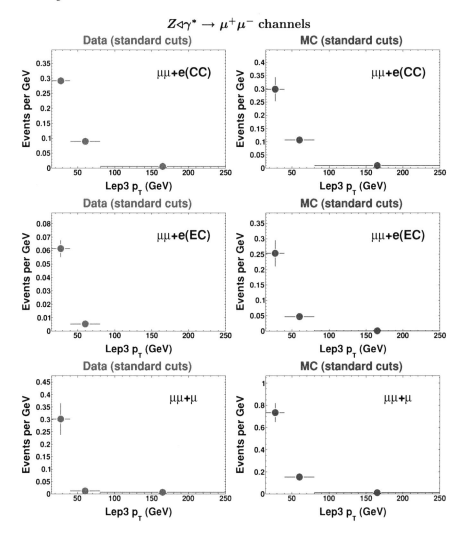

Fig. 7.44 *Left* matrix method estimate of the Z+jet background. *Right* MC prediction

7.11). The number of observed signal events is determined by allowing the expected signal yield to float so as to minimise the following log-likelihood test statistic:

$$Q = -2 \ln L = 2 \sum_{i=0}^{\text{bins}} \left[N_i^{\text{pred}} - N_i^{\text{obs}} + N_i^{\text{obs}} \ln \frac{N_i^{\text{obs}}}{N_i^{\text{pred}}} \right],$$

where N_i^{pred} is the predicted number of (signal plus background) events in bin i, and N_i^{obs} is the observed number of events. For $N_i^{\text{obs}} = 0$, we use $0 \ln 0 = 0$. The uncertainty on S_{sig} is given by the interval $\delta Q = \pm 1$.

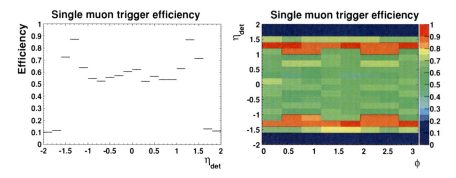

Fig. 7.45 Single muon trigger efficiencies with respect to medium offline quality

Table 7.9 Table of uncertainty sources in the ZZ/γ^* cross section measurement in the ee channel

	N_{bgd}^{pred}	N_{sig}^{pred}	N_{sig}^{obs}	A_{Z/γ^*}	A_{sig}	$A_{Z/\gamma^*}/A_{sig}$	σ_{sig} (pb)
Values	83.1	25.2	31.2	0.182	0.099	1.83	1.640
L_{inst} prof.	+1.1	+0.1	−0.5	+0.008	+0.005	−0.01	−0.030
z_{pv} prof.	−0.9	−0.1	+0.4	+0.002	+0.001	+0.01	+0.030
$Z/\gamma^* p_T$	−0.0	−0.0	+0.1	+0.000	+0.000	+0.00	+0.010
Diboson p_T	+2.2	+0.5	−0.6	+0.000	+0.002	−0.03	−0.060
JES	+0.8	+0.2	−0.5	+0.001	+0.001	−0.01	−0.030
JER	+0.7	−0.1	−0.5	−0.000	−0.000	+0.00	−0.020
ICR JES	−0.2	−0.1	+0.1	−0.000	−0.000	+0.00	+0.010
e p_T scale	−0.4	+0.1	+0.0	−0.000	+0.000	−0.00	+0.000
e p_T resol.	+0.3	+0.0	+0.0	−0.000	−0.000	−0.00	+0.000
μ p_T scale	+0.4	+0.1	−0.3	−0.000	+0.000	−0.00	−0.020
μ p_T resol.	+0.4	−0.2	−0.1	−0.000	−0.001	+0.02	+0.010
e p_T tails	−0.2	+0.0	+0.1	−0.001	−0.000	−0.00	+0.010
μ p_T tails	+0.5	−0.1	−0.7	−0.000	−0.001	+0.01	−0.030
Lep eff. versus p_T	+0.2	+0.1	−0.1	−0.000	+0.000	−0.01	−0.010
Lep eff. versus η	+0.5	+0.2	−0.1	−0.001	−0.000	−0.01	−0.010
Jet eff.	+0.9	−0.0	−0.4	+0.001	+0.000	+0.00	−0.020
Trkjet eff	−0.8	−0.1	+0.3	−0.001	−0.001	+0.01	+0.030
W+jets model.	+1.5	+0.0	−0.4	+0.000	+0.000	+0.00	−0.020
$W\gamma$ model.	+3.4	+0.0	−0.7	+0.000	+0.000	+0.00	−0.030
Total syst.	4.9	0.7	1.7	0.009	0.006	0.05	0.106
Stat.	0.0	0.0	8.5	0.000	0.000	0.00	0.450
Stat $\oplus$ syst.	4.9	0.7	8.7	0.009	0.006	0.05	0.462

7.14 Systematic Uncertainties

This analysis has been carefully designed to have minimal sensitivity to systematic uncertainties. Measurement of the signal cross sections as ratios to the inclu-

7.14 Systematic Uncertainties

Table 7.10 Table of uncertainty sources in the $WZ/\gamma^* \to \ell\nu\ell^+\ell^-$ cross section measurement in the (combining the four sub-channels)

	$N_{\text{bgd}}^{\text{pred}}$	$N_{\text{sig}}^{\text{pred}}$	$N_{\text{sig}}^{\text{obs}}$	A_{Z/γ^*}	A_{sig}	$A_{Z/\gamma^*}/A_{\text{sig}}$	σ_{sig} (pb)
Values	14.4	52.0	68.3	0.258	0.142	1.81	4.460
L_{inst} prof.	−0.5	+0.7	+0.7	+0.006	+0.006	−0.02	−0.010
z_{pv} prof.	−0.1	−0.2	+0.2	+0.003	+0.001	+0.01	+0.040
$Z/\gamma^* p_T$	+0.0	+0.0	+0.0	−0.000	−0.000	−0.00	+0.000
Diboson p_T	−0.0	+0.2	+0.0	+0.000	+0.000	−0.01	−0.010
JES	+0.8	−0.1	−0.6	−0.000	−0.000	+0.00	−0.030
JER	+0.3	−0.1	−0.4	+0.000	−0.000	+0.00	−0.010
ICR JES	−0.2	−0.0	+0.3	+0.000	+0.000	+0.00	+0.020
$e\ p_T$ scale	+0.1	+0.1	−0.1	−0.000	+0.000	−0.00	−0.010
$e\ p_T$ resol.	−0.2	−0.3	+1.0	+0.000	−0.000	+0.01	+0.090
$\mu\ p_T$ scale	−0.1	+0.1	+0.4	−0.000	+0.000	−0.00	+0.020
$\mu\ p_T$ resol.	+0.2	−0.0	+0.2	+0.000	+0.000	+0.00	+0.020
$e\ p_T$ tails	+0.7	+0.3	−0.1	−0.001	+0.000	−0.01	−0.030
$\mu\ p_T$ tails	+0.5	−0.2	−0.2	−0.000	−0.001	+0.01	+0.000
Trk eff.	−0.5	−0.3	+0.4	−0.001	−0.002	+0.01	+0.060
μ eff.	−0.5	−0.3	+0.7	−0.000	−0.001	+0.01	+0.070
e eff.	+0.0	−0.0	+0.0	−0.001	−0.000	+0.00	+0.000
Z+jets model.	+2.3	+0.0	−1.8	+0.000	+0.000	+0.00	−0.110
Total syst.	2.8	1.0	2.5	0.008	0.006	0.03	0.184
Stat.	0.0	0.0	9.4	0.000	0.000	0.00	0.610
Stat ⊕ syst.	2.8	1.0	9.7	0.008	0.006	0.03	0.637

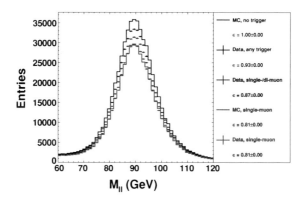

Fig. 7.46 Comparison of dimuon yields for different trigger requirements in data, and simulated trigger requirements in MC. The efficiencies are determined by comparing with the MC yield with no trigger simulation

sive $Z/\gamma^* \to \ell^+\ell^-$ cross section cancels the effect of many sources. For example, the luminosity uncertainty almost totally cancels. The ZZ/γ^* analysis is largely insensitive to uncertainties affecting lepton reconstruction, identification and trigger efficiencies. The WZ/γ^* analysis is however sensitive to these effects due to the additional lepton requirement, though less sensitive than a direct measurement of the absolute cross section. The normalisation to the inclusive $Z/\gamma^* \to \ell^+\ell^-$ yield

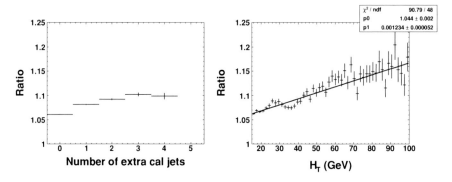

Fig. 7.47 Ratio of the dimuon yield in data with no trigger requirement to the yield with the requirement of a single-muon or dimuon trigger. The ratios are shown as a function of the number of jets, and H_T which is the p_T sum of the leading two jets

also means that we only need to rely on the MC event generators to predict the various background cross sections *relative* to the $Z/\gamma^* \to \ell^+\ell^-$ cross section. As we shall see, the systematic uncertainties are substantially smaller than the statistical uncertainties.

All sources of systematic uncertainty are evaluated with one-sided[5] variations on the simulation. Tables 7.9 and 7.10 list the effects of these variations on the cross section measurements for ZZ/γ^* and WZ/γ^* respectively. Further tables for the different sub-channels can be found in Appendix B. The shifts in the following quantities are provided: the number of predicted signal and background events; the number of observed signal events, the acceptances and acceptance ratio; the measured signal cross section. The following variations (only on the simulation unless otherwise stated) are considered:

- **Jet energy scale and resolution** Two variations are considered on all jets: scaling the jet momenta down by a factor of 1.1, and smearing the jet momenta by a Gaussian of width, $\delta E/E = 0.1$. An additional variation switches off the offset of -1 GeV on ICR jet energies (see Sect. 7.5). These variations are all propagated to the $\not{E}_T$.
- **Trackjet corrections** The trackjet efficiency correction (see Sect. 7.5) is switched off.
- **Lepton momentum scale** The magnitudes of the electron and muon momenta are scaled up by a factor of 1.002. These variations (and those on the resolution) are all propagated to the $\not{E}_T$.
- **Lepton momentum resolution (1)** Electron energies are smeared by a Gaussian of width $\delta E/E = 0.03$. Muon track curvatures are smeared by a Gaussian of width $\delta(1/p_T) = 0.001\,\text{GeV}^{-1}$.

---

[5] Since the systematic uncertainties are small compared to the statistical uncertainties, it is not considered necessary to evaluate two-sided variations. This saves a considerable amount of computing time.

7.14 Systematic Uncertainties

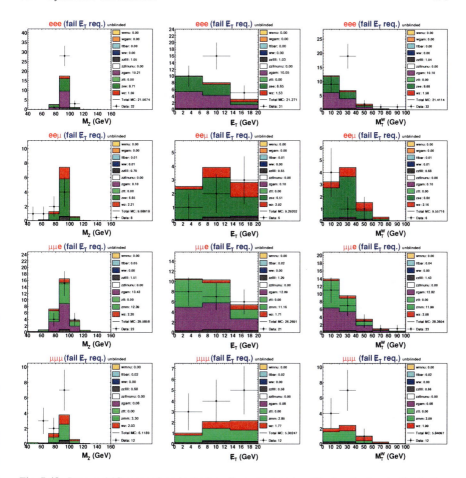

Fig. 7.48 Data-vs-MC comparison in the low-$\slashed{E}_T$ control region. The invariant mass distribution does not include the cut on that variable. The rows correspond to different sub-channels

- **Lepton momentum resolution (2)** For 5 % of electrons and muons, a significantly larger smearing is applied: $\delta E/E = 0.1$ for electrons, and $\delta(1/p_T) = 0.005$ for muons. The leakage of Drell-Yan events past the p'_T and q'_T cuts is potentially more sensitive to mis-modelling of such non-Gaussian tails, rather than the core Gaussian component.
- **Drell-Yan p_T modelling** The value of the g_2 parameter in ResBos is nominally set to the values measured in the ϕ^* shape measurement (see Chap. 6). We vary the value of g_2 by twice the uncertainties of that measurement. The factor of two accounts for the rather poor agreement with ResBos, and correspondingly large minimum χ^2 in the g_2 fits.
- **Diboson p_T modelling** The PYTHIA → POWHEG reweighting is switched off.

196 7 Measurement of the ZZ and WZ Production Cross Sections

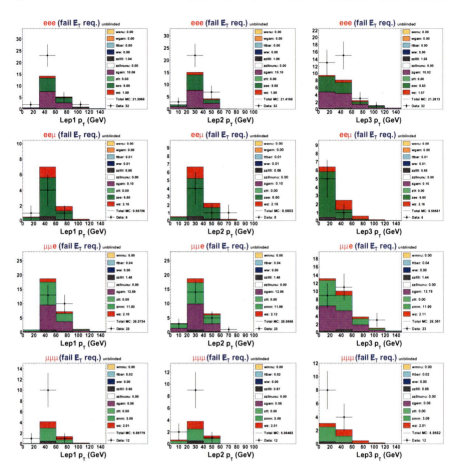

Fig. 7.49 Data-vs-MC comparison in the low-$\not{E}_T$ control region. The rows correspond to different sub-channels

- **Lepton ID efficiency** The efficiency corrections determined using the tag and probe method are varied within their statistical uncertainties. All bins are coherently shifted down. Since we are mostly interested in the absolute efficiency, this is a somewhat conservative approach, but nevertheless introduces no significant uncertainty.
- **Dependence of lepton efficiencies on p_T and η** This variation is only considered in the $ZZ/\gamma^* \to \nu\bar{\nu}\ell^+\ell^-$ analysis, which is relatively insensitive to the absolute lepton efficiencies. The first variation randomly throws away 15 % of leptons with $p_T < 25$ GeV. The second variation randomly throws away 15 % of leptons with $|\eta_{\text{det}}| > 1.7$.
- **Instantaneous luminosity profile** The instantaneous luminosity profile reweighting is switched off. The scale-factors are derived using the same MC luminos-

7.14 Systematic Uncertainties

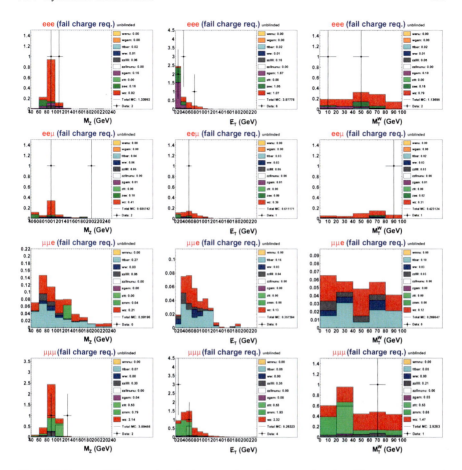

Fig. 7.50 Data-vs-MC comparison in the same charge sign control region. The rows correspond to different sub-channels

ity profile as in the cross section analysis. This variation therefore conservatively accounts for possible systematic biases in the measurement of lepton ID efficiency scale-factors.

- **W+jets normalisation** The W+jets normalisation is varied within the statistical uncertainties of the matrix method based estimate. All other sources of systematic uncertainty are allowed to vary the shape of the this background as predicted by the MC.
- **Z+jets normalisation** The Z+jets normalisation is varied within the statistical uncertainties of the matrix method based estimate. All other sources of systematic uncertainty are allowed to vary the shape of the this background as predicted by the MC.

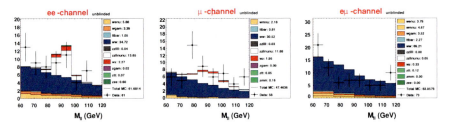

Fig. 7.51 Dilepton invariant mass distribution of the selected signal candidate events

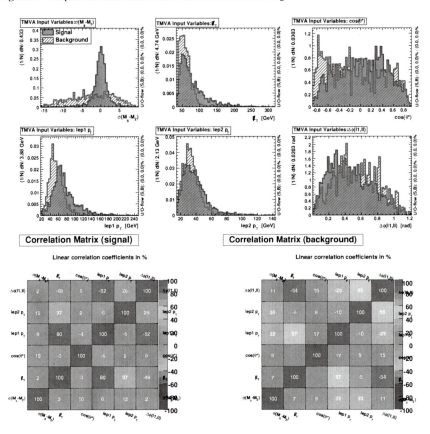

Fig. 7.52 *Top two rows* comparison of signal and background in the *ee* channel for each of the MVA training variables. *Bottom row* training variable correlation matrices for (*left*) signal and (*right*) background

- **$W\gamma$ cross section** The $W\gamma$ normalisation is varied by a factor of two to account for the fact that PYTHIA does not include the matrix element for wide angle final state photon emission (Figs. 7.58, 7.59, 7.60).

7.15 Results

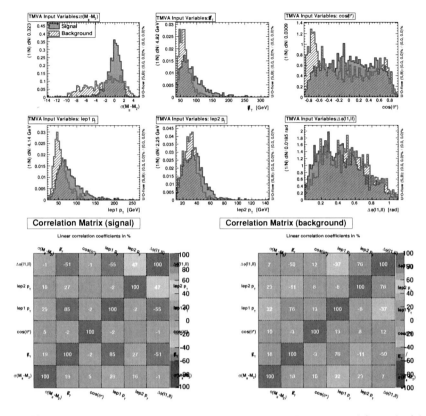

Fig. 7.53 *Top two rows* comparison of signal and background in the $\mu\mu$ channel for each of the MVA training variables. *Bottom row* training variable correlation matrices for (*left*) signal and (*right*) background

7.15 Results

Standard Model predictions are calculated using the NLO program MCFM [17] with the MSTW2008 NLO PDFs [18]. The renormalisation and factorisation scales are set to $2m_Z$ and $m_W + m_Z$ for the ZZ/γ^* and WZ/γ^* predictions respectively. For $Z/\gamma^* \to \ell^+\ell^-$ invariant masses between 60 and 130 GeV, the predictions are $\sigma(ZZ/\gamma^*) = 1.3 \pm 0.1$ pb and $\sigma(WZ/\gamma^*) = 3.2 \pm 0.2$ pb. The uncertainties correspond to variations of the renormalisation and factorisation scales by a factor of two. For $Z/\gamma^* \to \ell^+\ell^-$ invariant masses between 60 and 130 GeV, the measured signal cross sections are:

$$\sigma(p\bar{p} \to ZZ/\gamma^*) = 1.64 \pm 0.46 \, \text{pb},$$
$$\sigma(p\bar{p} \to WZ/\gamma^*) = 4.46 \pm 0.64 \, \text{pb}.$$

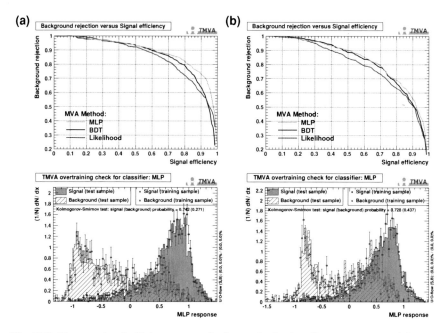

Fig. 7.54 *Top row* signal efficiency versus background rejection. *Bottom row* overtraining test distributions for the MLP method

These are both in agreement with the Standard Model predictions. Figure 7.61 compares measured cross sections from this and other analyses[6] with the Standard Model predictions from MCFM. The individual measurements for each sub-channel in this analysis are also presented and are in good agreement with each other.

---

[6] It should be noted that some of the other measurements are translated into pure ZZ or WZ cross sections. The previous D0 analysis of the $ZZ/\gamma^* \to \nu\bar{\nu}\ell^+\ell^-$ process used the MCFM [17]
(Footnote 6 continued)
program to estimate a correction factor of 3.4 % that converts a ZZ/γ^* cross section into a pure ZZ/γ^* cross section. Considering the overall uncertainties, such details do not significantly affect the comparison with previous measurements. These previous measurements are also discussed in Chap. 1.

7.15 Results

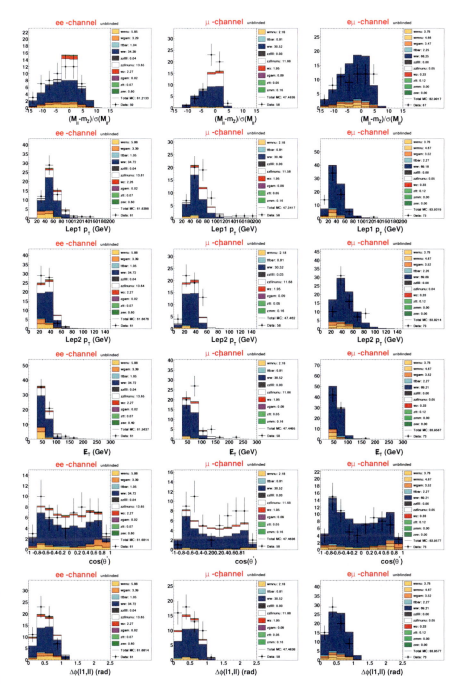

Fig. 7.55 Distribution of the kinematic variables that best discriminate between WW and $ZZ/\gamma^* \to \nu\bar{\nu}\ell^+\ell^-$ for the signal candidates after *all* selection cuts

202   7 Measurement of the ZZ and WZ Production Cross Sections

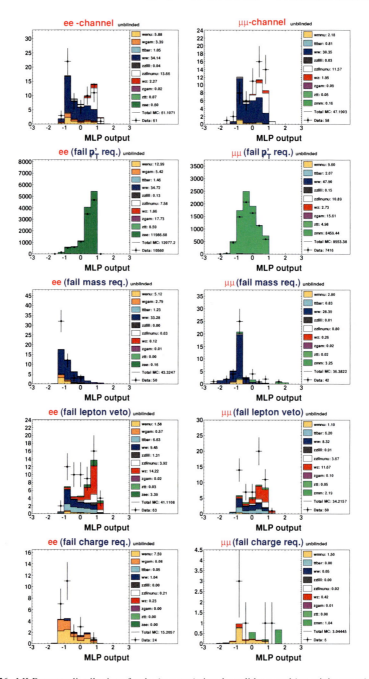

Fig. 7.56 MLP output distributions for the (*top row*) signal candidates and (remaining rows) signal-free control regions

7.15 Results

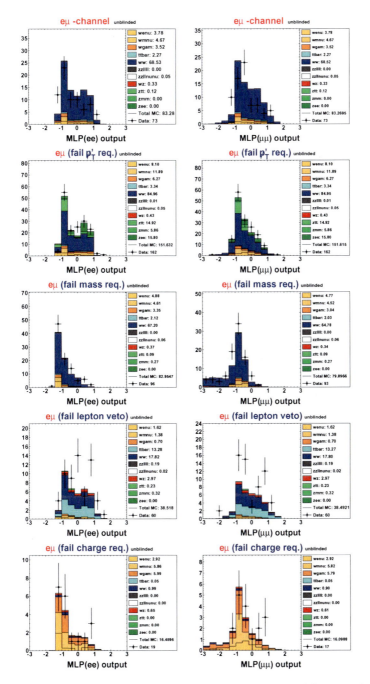

Fig. 7.57 MLP output distributions for the (*top row*) signal region and (remaining rows) signal-free control regions in the $e\mu$ channels. The *left-* and *right-hand columns* correspond to the ee and $\mu\mu$ versions of the MLP respectively

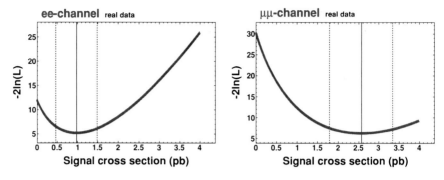

Fig. 7.58 Variation of the log-likelihood test statistic with the ZZ/γ^* signal cross section

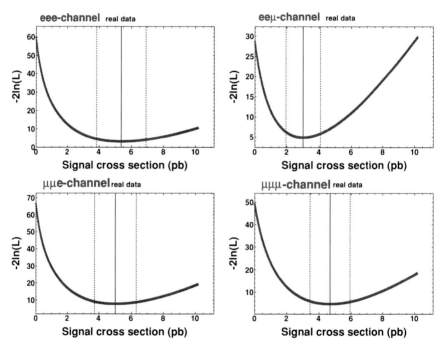

Fig. 7.59 Variation of the log-likelihood test statistic with the WZ/γ^* signal cross section

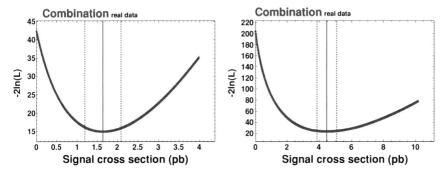

Fig. 7.60 Variation of the log-likelihood test statistic with the (*left*) ZZ/γ^* and (*right*) WZ/γ^* signal cross sections after combining all sub-channels

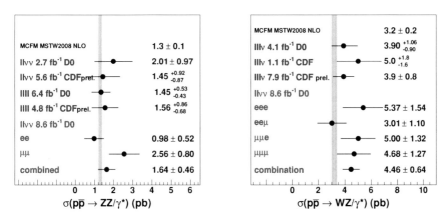

Fig. 7.61 Comparison of the (*left*) WZ/γ^* and (*right*) ZZ/γ^* production cross sections with SM predictions and with previous experimental measurements. The quoted uncertainties combine systematic and statistical components

References

1. T. Sjostrand, Comput. Phys. Commun. **135**, 238 (2001)
2. M.L. Mangano, JHEP **07**, 001 (2003)
3. V.M. Abazov et al., Phys. Rev. D **78**, 072002 (2008)
4. V.M. Abazov et al., Phys. Lett. B **695**, 67 (2011)
5. C. Balazs, C.-P. Yuan, Phys. Rev. D **56**, 5558–5583 (1997)
6. P. Nason, JHEP **0411**, 40 (2004)
7. S. Frixione, P. Nason, C. Oleari, JHEP **0711**, 70 (2007)
8. K. DeVaughan et al., D0 Note 5801, D0, 2008
9. M. Cooke et al., D0 Note 6031, D0, 2010
10. B. Calpas, J. Kraus, T. Yasuda, D0 Note 6051, D0, 2010
11. M. Vesterinen, D0 Note 6018, D0, 2011
12. K. Nakamura et al., J. Phys. G **37**, 075021 (2010)
13. A. Hoecker, P. Speckmayer, J. Stelzer, J. Therhaag, E. von Toerne, H. Voss, CERN-OPEN 007, 2007

14. J.D. Hobbs, T. Nunnemann, R. Van Kooten, Study of $p\bar{p} \to z/\gamma^* \to ee$ and $\mu\mu$ event yields as a luminosity cross check. D0 note 5268, D0, 2006
15. R. Hamburg, W.L. van Neerven, T. Matsuura, Nucl. Phys. B **359**, 343 (1991)
16. A.D. Martin, R.G. Roberts, W.J. Stirling, R.S. Thorne, Phys. Lett. B **604**, 61 (2004)
17. J.M. Campbell, R.K. Ellis, Phys. Rev. D **60**, 113006 (2009)
18. A. Martin, W. Stirling, R. Thorne, G. Watt, Parton distributions for the lhc. Eur. Phys. J. C **63**, 189–285 (2009)

Chapter 8
Conclusions

The Drell-Yan ($p\bar{p}/pp \to Z/\gamma^* \to \ell^+\ell^-$) process offers a unique insight into Quantum chromodynamics (QCD). Higher order effects in QCD directly affect the dilepton transverse momentum, p_T, distribution. It is interesting to verify that perturbative calculations are able to accurately describe the shape of this distribution. In addition, the very low p_T region is sensitive to non-perturbative effects that cannot be calculated from first principles. The dependence of these non-perturbative effects on the dilepton rapidity is rather poorly constrained by existing data. The Tevatron has collected huge samples of dilepton events with an invariant mass around the Z boson mass. Despite the available statistics, the p_T distribution can only be measured with limited precision due to the large corrections that are needed for experimental resolution and efficiency. Measurements with only one tenth of the final dataset were already limited by systematic uncertainties.

Here, alternative observables are proposed, that are sensitive to the relevant physics but are less susceptible to detector effects. The a_T variable is one component of the p_T, with respect to a particular reference axis. For $\Delta\phi > \pi/2$, where $\Delta\phi$ is the azimuthal opening angle between the leptons, a_T has significantly better resolution than the p_T. Compared to the p_T, the a_T component is also significantly less correlated with the efficiency of selection requirements usually imposed for analyses of the $Z/\gamma^* \to \ell^+\ell^-$ final state. Unfortunately, the resolution of a_T rapidly degrades at smaller $\Delta\phi$ (corresponding to higher p_T). A simple improvement is to divide a_T by the measured dilepton invariant mass, $M_{\ell\ell}$. Mis-measurement in the p_T of one or both of the leptons leads to a correlated change in a_T and $M_{\ell\ell}$, that is partially cancelled in the ratio. Compared to the a_T itself, the ratio $a_T/M_{\ell\ell}$ is found to be no less sensitive to parameters describing the shape of the p_T distribution. To some extent, this idea also works for the ratio of p_T over $M_{\ell\ell}$, which has slightly better resolution than the p_T alone. In order to achieve the best possible resolution, one needs a variable that is constructed exclusively from angles, e.g., the $\Delta\phi$. In fact, the $\Delta\phi$ is primarily sensitive to the a_T component, and thus also benefits from the reduced efficiency dependence. However, the translation from $\Delta\phi$ to a_T also depends on the scattering angle in the dilepton rest frame, θ^*. This degrades the sensitivity of $\Delta\phi$ to the physics of interest. The following variable is proposed as a simple modification of the $\Delta\phi$:

$$\phi_\eta^* \equiv \tan(\phi_{\text{acop}}/2)\sin\theta^*,$$

where $\phi_{\text{acop}} = \pi - \Delta\phi$. An approximation for the scattering angle that relies only on the lepton angles, is,

$$\cos\theta_\eta^* = \tanh\left(\frac{\eta^- - \eta^+}{2}\right),$$

with η^- and η^+ being the pseudorapidities of the negatively and positively charged leptons respectively. It is demonstrated that ϕ_η^* has better sensitivity to the p_T physics than $\Delta\phi$, and is still determined exclusively from angles, therefore having essentially perfect experimental resolution considering the bin sizes that are feasible, even with the huge $Z/\gamma^* \to \ell^+\ell^-$ event samples.

Using 7.3 fb^{-1} of data in the ee and $\mu\mu$ decay channels, we measure the shape of the Drell-Yan ϕ_η^* distribution in the region $70 < M_{\ell\ell} < 110\,\text{GeV}$, and in three bins of dilepton rapidity. One of the challenges is to accurately model the lepton identification and trigger efficiencies in poorly instrumented regions of the detector that are back-to-back in ϕ. Thanks to the large event samples, and the relative insensitivity of ϕ_η^* to detector effects, this measurement is significantly more precise than any previous measurements of the p_T distribution. A prediction from the RESBOS program is unable to describe the shape of the distribution in detail. In particular, a modified non-perturbative form factor that increases the width of the p_T distribution at large rapidities ($|y| > 2$) is strongly disfavoured.

Electroweak boson pair production processes provide a window into the non-abelian gauge structure of the Standard Model through their sensitivity to triple gauge couplings. We study the production of WZ/γ^* and ZZ/γ^*, which are the two diboson processes in the Standard Model with the lowest cross section, aside from the associated production of the Higgs boson with a W or Z. The $ZZ/\gamma^* \to \nu\bar{\nu}\ell^+\ell^-$ and $WZ/\gamma^* \to \ell\nu\ell^+\ell^-$ final states are analysed. A key challenge is to separate ZZ/γ^* signal events with genuine missing transverse momentum, from mis-reconstructed $Z/\gamma^* \to \ell^+\ell^-$ events. Taking into account systematic uncertainties, in order to achieve the best possible precision on the signal cross section, a surprisingly large $Z/\gamma^* \to \ell^+\ell^-$ rejection of around 10^6 is required. A very conservative approach is therefore needed in the construction of the discriminating variables. The measured signal cross sections are, for $Z/\gamma^* \to \ell^+\ell^-$ invariant masses between 60 and 130 GeV,

$$\sigma(p\bar{p} \to ZZ/\gamma^*) = 1.64 \pm 0.46\,\text{pb},$$

$$\sigma(p\bar{p} \to WZ/\gamma^*) = 4.46 \pm 0.64\,\text{pb}.$$

8 Conclusions

These are in reasonable agreement with Standard Model predictions, at next-to-leading-order accuracy, of 1.3 ± 0.1 pb and 3.2 ± 0.2 pb for the ZZ/γ^* and WZ/γ^* cross sections respectively.

Looking to the future, the emphasis moves to the LHC for the study of electroweak physics. The ϕ_η^* variable could potentially improve studies of $Z/\gamma^* \to \ell^+\ell^-$ production. Although the LHC detectors have significantly better resolution, the available event samples will eventually be significantly larger than those at the Tevatron. Precision measurements of $Z/\gamma^* \to \ell^+\ell^-$ production will therefore rapidly become limited by systematic uncertainties, to which variables like ϕ_η^* are less susceptible. The electroweak diboson production cross sections are substantially larger at the LHC. In the near future, this will allow precision measurements of the production and decay properties. Some of the challenges will be common to this analysis, e.g., rejecting mis-reconstructed Drell-Yan events in the study of $ZZ/\gamma^* \to \nu\bar{\nu}\ell^+\ell^-$ production. The LHC analyses may therefore benefit from some of the ideas presented in this work.

Appendix A
Electron and Muon Transfer Functions

Figure A.1 shows the variation of the curvature resolution $\delta(1p_T)$ as a function of p_T for different categories of muons. The curvature resolution is parameterised as

$$\delta(1/p_T) = A + B/p_T.$$

Figure A.2 shows the variation of the mean $\delta E/E$ with E for electrons as predicted by the MC simulation. Electrons are categorised based on the region of the calorimeter in which they are reconstructed. The energy resolution is parameterised as

$$\delta(E)/E = C + S/E^{1/2} + N/E.$$

For type-2 IC electrons, the central track is used for kinematic analysis, so the track curvature resolution is shown instead, with the same parameterisation as for muons.

Appendix A: Electron and Muon Transfer Functions

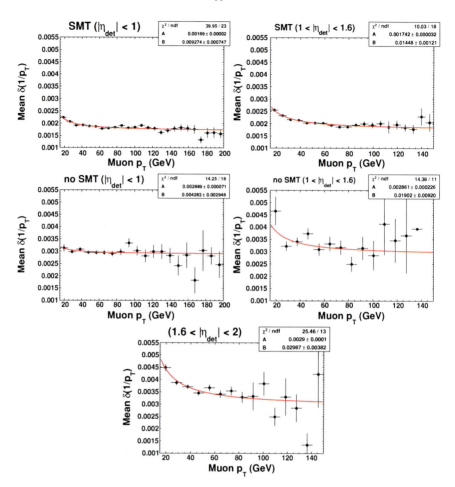

Fig. A.1 Muon p_T transfer functions

Appendix A: Electron and Muon Transfer Functions 213

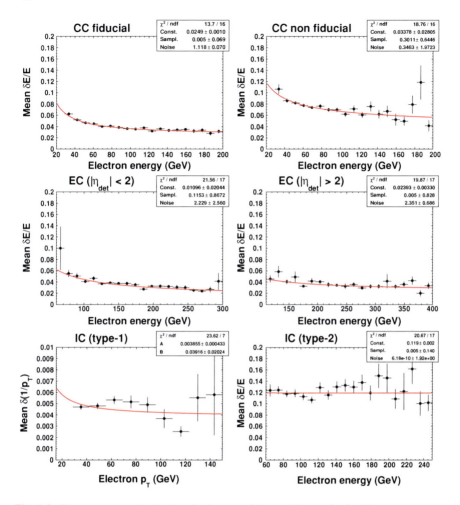

Fig. A.2 Electron energy (momentum in the case of type-1 IC) transfer functions

Appendix B
Systematic Uncertainty Tables by Sub-Channel

Tables B.1–B.6 list the effect of each source of systematic uncertainty on the signal cross sections in the different sub-channels of the ZZ/γ^* and WZ/γ^* analyses.

Table B.1 Table of uncertainty sources in the ZZ/γ^* cross section measurement in the ee channel

	N_{bgd}^{pred}	N_{sig}^{pred}	N_{sig}^{obs}	A_{Z/γ^*}	A_{sig}	$A_{Z/\gamma^*}/A_{sig}$	σ_{sig} (pb)
Values	47.4	13.7	10.0	0.178	0.109	1.64	0.980
L_{inst} prof.	−0.0	+0.2	+0.2	+0.006	+0.005	−0.02	+0.000
z_{pv} prof.	−0.7	−0.1	+0.3	+0.003	+0.001	+0.01	+0.030
$Z/\gamma^* p_T$	−0.0	−0.0	+0.0	+0.000	+0.000	+0.00	+0.000
Diboson p_T	+1.1	+0.2	−0.1	+0.000	+0.002	−0.03	−0.030
JES	+0.3	+0.1	+0.0	+0.002	+0.002	−0.01	−0.010
JER	+0.2	−0.0	−0.1	−0.000	−0.000	+0.00	−0.020
ICR JES	+0.0	−0.0	+0.1	−0.001	−0.000	+0.00	+0.000
$e\ p_T$ scale	−0.4	+0.1	+0.1	−0.001	+0.000	−0.01	+0.000
$e\ p_T$ resol.	−0.0	+0.0	+0.1	−0.001	−0.000	−0.00	+0.000
$\mu\ p_T$ scale	+0.0	+0.0	+0.0	+0.000	+0.000	+0.00	+0.000
$\mu\ p_T$ resol.	+0.1	+0.0	−0.1	−0.001	+0.000	−0.01	−0.020
$e\ p_T$ tails	−0.5	−0.0	+0.2	−0.001	−0.001	+0.00	+0.010
$\mu\ p_T$ tails	−0.0	+0.0	+0.1	−0.000	−0.000	−0.00	+0.000
Lep eff. versus p_T	+0.1	+0.1	+0.1	−0.001	+0.000	−0.01	+0.000
Lep eff. versus η	+0.2	+0.1	+0.2	−0.001	−0.000	−0.01	+0.000
Jet eff.	+0.3	−0.0	−0.1	+0.001	+0.000	+0.00	−0.010
Trkjet eff	−0.5	−0.1	+0.3	−0.001	−0.001	+0.01	+0.030
W+jets model.	+1.2	+0.0	−0.2	+0.000	+0.000	+0.00	−0.020
$W\gamma$ model.	+3.4	+0.0	−0.6	+0.000	+0.000	+0.00	−0.060
Total syst.	4.0	0.4	0.9	0.007	0.006	0.04	0.088
Stat.	0.0	0.0	5.3	0.000	0.000	0.00	0.510
Stat ⊕ syst.	4.0	0.4	5.4	0.007	0.006	0.04	0.518

M. Vesterinen, *Z Boson Transverse Momentum Distribution, and ZZ and WZ Production*, Springer Theses, DOI: 10.1007/978-3-642-30788-1,
© Springer-Verlag Berlin Heidelberg 2012

Table B.2 Table of uncertainty sources in the ZZ/γ^* cross section measurement in the $\mu\mu$ channel

Values	$N_{\text{bgd}}^{\text{pred}}$	$N_{\text{sig}}^{\text{pred}}$	$N_{\text{sig}}^{\text{obs}}$	A_{Z/γ^*}	A_{sig}	$A_{Z/\gamma^*}/A_{\text{sig}}$	σ_{sig} (pb)
	35.6	11.6	22.3	0.186	0.090	2.07	2.560
L_{inst} prof.	+1.1	−0.0	−0.5	+0.011	+0.005	+0.00	−0.050
z_{pv} prof.	−0.2	−0.0	+0.3	+0.002	+0.001	+0.01	+0.040
$Z/\gamma^* p_T$	−0.0	−0.0	+0.0	+0.000	+0.000	+0.00	+0.000
Diboson p_T	+1.1	+0.2	−0.5	+0.000	+0.002	−0.04	−0.100
JES	+0.5	+0.1	−0.3	+0.001	+0.001	−0.02	−0.060
JER	+0.5	−0.0	−0.2	−0.000	−0.000	+0.01	−0.020
ICR JES	−0.2	−0.0	+0.2	−0.000	−0.000	+0.01	+0.030
$e\ p_T$ scale	−0.0	+0.0	+0.0	+0.000	+0.000	+0.00	+0.000
$e\ p_T$ resol.	+0.3	−0.0	−0.1	+0.000	+0.000	+0.00	−0.010
$\mu\ p_T$ scale	+0.4	+0.1	−0.3	−0.001	+0.000	−0.01	−0.040
$\mu\ p_T$ resol.	+0.4	−0.3	+0.2	+0.000	−0.002	+0.05	+0.080
$e\ p_T$ tails	+0.4	+0.1	−0.1	−0.001	−0.000	−0.01	−0.020
$\mu\ p_T$ tails	+0.5	−0.1	−0.5	−0.001	−0.001	+0.03	−0.030
Lep eff. versus p_T	+0.1	+0.0	+0.0	−0.000	+0.000	−0.00	+0.000
Lep eff. versus η	+0.4	+0.1	−0.1	−0.001	+0.000	−0.01	−0.030
Jet eff.	+0.6	+0.0	−0.2	+0.001	+0.000	−0.00	−0.030
Trkjet eff	−0.2	−0.1	+0.1	−0.001	−0.001	+0.01	+0.020
W+jets model.	+0.3	+0.0	−0.1	+0.000	+0.000	+0.00	−0.010
$W\gamma$ model.	+0.0	+0.0	+0.0	+0.000	+0.000	+0.00	+0.000
Total syst.	2.1	0.4	1.1	0.011	0.006	0.08	0.175
Stat.	0.0	0.0	6.8	0.000	0.000	0.00	0.780
Stat $\oplus$ syst.	2.1	0.4	6.9	0.011	0.006	0.08	0.799

Appendix B: Systematic Uncertainty Tables by Sub-Channel 217

Table B.3 Table of uncertainty sources in the WZ/γ^* cross section measurement in the eee channel

	$N_{\text{bgd}}^{\text{pred}}$	$N_{\text{sig}}^{\text{pred}}$	$N_{\text{sig}}^{\text{obs}}$	A_{Z/γ^*}	A_{sig}	$A_{Z/\gamma^*}/A_{\text{sig}}$	σ_{sig} (pb)
Values	1.6	9.8	15.5	0.248	0.107	2.31	5.370
L_{inst} prof.	−0.1	+0.1	+0.1	+0.004	+0.002	−0.01	+0.000
z_{pv} prof.	−0.0	−0.0	+0.1	+0.004	+0.001	+0.01	+0.030
$Z/\gamma^* p_T$	+0.0	+0.0	+0.0	−0.000	+0.000	−0.00	+0.000
Diboson p_T	−0.0	+0.0	+0.0	+0.000	+0.000	−0.00	+0.000
JES	+0.3	−0.0	−0.2	+0.000	−0.000	+0.00	−0.090
JER	+0.1	−0.0	−0.1	+0.000	−0.000	+0.01	−0.020
ICR JES	−0.0	+0.0	+0.1	+0.000	+0.000	−0.00	+0.020
$e\,p_T$ scale	+0.1	+0.0	+0.0	−0.001	−0.000	−0.01	−0.040
$e\,p_T$ resol.	+0.1	−0.1	−0.1	+0.000	−0.000	+0.01	+0.000
$\mu\,p_T$ scale	+0.0	+0.0	+0.0	+0.000	+0.000	+0.00	+0.000
$\mu\,p_T$ resol.	−0.0	+0.0	+0.0	−0.000	+0.000	−0.00	+0.000
$e\,p_T$ tails	+0.1	+0.1	−0.1	−0.001	+0.000	−0.01	−0.070
$\mu\,p_T$ tails	−0.1	+0.0	+0.1	−0.000	+0.000	−0.00	+0.020
Trk eff.	−0.1	−0.0	+0.1	−0.002	−0.002	+0.01	+0.030
μ eff.	+0.0	+0.0	+0.0	+0.000	+0.000	+0.00	+0.000
e eff.	−0.0	−0.0	+0.0	−0.001	−0.000	+0.00	+0.000
Z+jets model.	+0.1	+0.0	+0.0	+0.000	+0.000	+0.00	−0.020
Total syst.	0.4	0.1	0.3	0.006	0.003	0.03	0.134
Stat.	0.0	0.0	4.4	0.000	0.000	0.00	1.530
Stat $\oplus$ syst.	0.4	0.1	4.4	0.006	0.003	0.03	1.536

Table B.4 Table of uncertainty sources in the WZ/γ^* cross section measurement in the $ee\mu$ channel

	$N_{\text{bgd}}^{\text{pred}}$	$N_{\text{sig}}^{\text{pred}}$	$N_{\text{sig}}^{\text{obs}}$	A_{Z/γ^*}	A_{sig}	$A_{Z/\gamma^*}/A_{\text{sig}}$	σ_{sig} (pb)
Values	1.6	9.8	15.5	0.248	0.107	2.31	5.370
L_{inst} prof.	−0.1	+0.1	+0.1	+0.004	+0.002	−0.01	+0.000
z_{pv} prof.	−0.0	−0.0	+0.1	+0.004	+0.001	+0.01	+0.030
$Z/\gamma^* p_T$	+0.0	+0.0	+0.0	−0.000	+0.000	−0.00	+0.000
Diboson p_T	−0.0	+0.0	+0.0	+0.000	+0.000	−0.00	+0.000
JES	+0.3	−0.0	−0.2	+0.000	−0.000	+0.00	−0.090
JER	+0.1	−0.0	−0.1	+0.000	−0.000	+0.01	−0.020
ICR JES	−0.0	+0.0	+0.1	+0.000	+0.000	−0.00	+0.020
$e\,p_T$ scale	+0.1	+0.0	+0.0	−0.001	−0.000	−0.01	−0.040
$e\,p_T$ resol.	+0.1	−0.1	−0.1	+0.000	−0.000	+0.01	+0.000
$\mu\,p_T$ scale	+0.0	+0.0	+0.0	+0.000	+0.000	+0.00	+0.000
$\mu\,p_T$ resol.	−0.0	+0.0	+0.0	−0.000	+0.000	−0.00	+0.000
$e\,p_T$ tails	+0.1	+0.1	−0.1	−0.001	+0.000	−0.01	−0.070
$\mu\,p_T$ tails	−0.1	+0.0	+0.1	−0.000	+0.000	−0.00	+0.020
Trk eff.	−0.1	−0.0	+0.1	−0.002	−0.002	+0.01	+0.030
μ eff.	+0.0	+0.0	+0.0	+0.000	+0.000	+0.00	+0.000
e eff.	−0.0	−0.0	+0.0	−0.001	−0.000	+0.00	+0.000
Z+jets model.	+0.1	+0.0	+0.0	+0.000	+0.000	+0.00	−0.020
Total syst.	0.4	0.1	0.3	0.006	0.003	0.03	0.134
Stat.	0.0	0.0	4.4	0.000	0.000	0.00	1.530
Stat $\oplus$ syst.	0.4	0.1	4.4	0.006	0.003	0.03	1.536

Table B.5 Table of uncertainty sources in the WZ/γ^* cross section measurement in the $\mu\mu e$ channel

	$N_{\text{bgd}}^{\text{pred}}$	$N_{\text{sig}}^{\text{pred}}$	$N_{\text{sig}}^{\text{obs}}$	A_{Z/γ^*}	A_{sig}	$A_{Z/\gamma^*}/A_{\text{sig}}$	σ_{sig} (pb)
Values	4.2	13.9	20.5	0.268	0.153	1.75	5.000
L_{inst} prof.	−0.2	+0.1	+0.2	+0.009	+0.006	−0.01	+0.020
z_{pv} prof.	−0.1	−0.0	+0.1	+0.003	+0.001	+0.00	+0.040
$Z/\gamma^* p_T$	+0.0	−0.0	+0.0	+0.000	+0.000	+0.00	+0.000
Diboson p_T	−0.0	+0.0	+0.0	+0.000	+0.000	−0.00	+0.000
JES	+0.6	−0.1	−0.4	−0.001	−0.001	+0.01	−0.060
JER	+0.2	−0.0	−0.3	+0.000	−0.000	+0.00	−0.050
ICR JES	−0.1	+0.0	+0.1	+0.000	+0.000	−0.00	+0.040
$e\, p_T$ scale	+0.0	+0.0	+0.0	+0.000	+0.000	−0.00	−0.010
$e\, p_T$ resol.	−0.3	−0.0	+0.6	+0.001	+0.000	+0.00	+0.160
$\mu\, p_T$ scale	−0.2	+0.1	+0.2	−0.001	+0.000	−0.01	+0.040
$\mu\, p_T$ resol.	−0.1	−0.1	+0.1	+0.001	−0.001	+0.01	+0.070
$e\, p_T$ tails	+0.1	+0.0	+0.0	−0.001	−0.000	−0.00	−0.010
$\mu\, p_T$ tails	+0.3	−0.1	−0.4	−0.000	−0.001	+0.01	−0.060
Trk eff.	−0.2	−0.1	+0.1	−0.000	−0.002	+0.02	+0.090
μ eff.	−0.2	−0.1	+0.3	−0.000	−0.001	+0.01	+0.110
e eff.	+0.0	−0.1	+0.0	+0.000	−0.001	+0.01	+0.020
Z+jets model.	+0.2	+0.0	−0.3	+0.000	+0.000	+0.00	−0.060
Total syst.	0.9	0.3	1.0	0.010	0.007	0.03	0.264
Stat.	0.0	0.0	5.3	0.000	0.000	0.00	1.290
Stat ⊕ syst.	0.9	0.3	5.4	0.010	0.007	0.03	1.317

Table B.6 Table of uncertainty sources in the WZ/γ^* cross section measurement in the $\mu\mu\mu$ channel

	$N_{\text{bgd}}^{\text{pred}}$	$N_{\text{sig}}^{\text{pred}}$	$N_{\text{sig}}^{\text{obs}}$	A_{Z/γ^*}	A_{sig}	$A_{Z/\gamma^*}/A_{\text{sig}}$	σ_{sig} (pb)
Values	4.7	14.5	20.0	0.268	0.160	1.68	4.680
L_{inst} prof.	−0.3	+0.3	+0.2	+0.009	+0.008	−0.03	−0.030
z_{pv} prof.	−0.1	−0.1	+0.0	+0.003	+0.001	+0.01	+0.040
$Z/\gamma^* p_T$	+0.0	−0.0	+0.0	+0.000	+0.000	+0.00	+0.000
Diboson p_T	−0.0	+0.1	−0.1	+0.000	+0.001	−0.01	−0.030
JES	−0.1	−0.0	+0.0	−0.001	−0.001	+0.00	+0.020
JER	−0.1	−0.0	+0.1	+0.000	−0.000	+0.00	+0.040
ICR JES	−0.0	−0.0	+0.0	+0.000	−0.000	+0.00	+0.000
$e\, p_T$ scale	+0.0	+0.0	+0.0	+0.000	+0.000	+0.00	+0.000
$e\, p_T$ resol.	−0.0	−0.0	+0.2	+0.001	+0.000	+0.00	+0.070
$\mu\, p_T$ scale	+0.0	+0.0	+0.2	−0.001	−0.000	−0.00	+0.040
$\mu\, p_T$ resol.	+0.2	+0.0	−0.1	+0.001	+0.001	−0.00	−0.030
$e\, p_T$ tails	−0.0	+0.1	+0.2	−0.001	+0.000	−0.01	+0.040
$\mu\, p_T$ tails	+0.2	−0.1	−0.1	−0.000	−0.001	+0.01	+0.020
Trk eff.	−0.2	−0.1	+0.0	−0.000	−0.001	+0.01	+0.040
μ eff.	−0.4	−0.1	+0.5	−0.000	−0.002	+0.02	+0.160
e eff.	+0.0	+0.0	+0.0	+0.000	+0.000	+0.00	+0.000
Z+jets model.	+1.1	+0.0	−0.9	+0.000	+0.000	+0.00	−0.200
Total syst.	1.3	0.3	1.1	0.010	0.009	0.04	0.286
Stat.	0.0	0.0	5.3	0.000	0.000	0.00	1.240
Stat ⊕ syst.	1.3	0.3	5.4	0.010	0.009	0.04	1.273